Mold Design Using NX 11.0: A Tutorial Approach

CADCIM Technologies
525 St. Andrews Drive
Schererville, IN 46375, USA
(www.cadcim.com)

Contributing Author

Prof. Sham Tickoo
Purdue University Northwest
Department of Mechanical Engineering Technology
Hammond, Indiana, USA

CADCIM
Technologies
Excellence in Technology

CADCIM Technologies

Mold Design Using NX 11.0: A Tutorial Approach
Sham Tickoo

CADCIM Technologies
525 St Andrews Drive
Schererville, Indiana 46375, USA
www.cadcim.com

ISBN 978-1-942689-81-2

www.cadcim.com

DEDICATION

*To teachers, who make it possible to disseminate knowledge
to enlighten the young and curious minds
of our future generations*

*To students, who are dedicated to learning new technologies
and making the world a better place to live in*

THANKS

*To the faculty and students of the MET department of
Purdue University Northwest for their cooperation*

To employees at CADCIM Technologies for their valuable help

Online Training Program Offered by CADCIM Technologies

CADCIM Technologies provides effective and affordable virtual online training on various software packages including Computer Aided Design, Manufacturing, and Engineering (CAD/CAM/CAE), computer programming languages, animation, architecture, and GIS. The training is delivered 'live' via Internet at any time, any place, and at any pace to individuals as well as the students of colleges, universities, and CAD/CAM training centers. The main features of this program are:

Training for Students and Companies in a Classroom Setting

Highly experienced instructors and qualified engineers at CADCIM Technologies conduct the classes under the guidance of Prof. Sham Tickoo of Purdue University Northwest, USA. This team has authored several textbooks that are rated "one of the best" in their categories and are used in various colleges, universities, and training centers in North America, Europe, and in other parts of the world.

Training for Individuals

CADCIM Technologies with its cost effective and time saving initiative strives to deliver the training in the comfort of your home or work place, thereby relieving you from the hassles of traveling to training centers.

Training Offered on Software Packages

CADCIM provides basic and advanced training on the following software packages:

CAD/CAM/CAE*: CATIA, Pro/ENGINEER Wildfire, Creo Parametric, Creo Direct, SolidWorks, Autodesk Inventor, Solid Edge, NX, AutoCAD, AutoCAD LT, AutoCAD Plant 3D, Customizing AutoCAD, EdgeCAM, and ANSYS*

Architecture and GIS*: Autodesk Revit Architecture, AutoCAD Civil 3D, Autodesk Revit Structure, AutoCAD Map 3D, Revit MEP, Navisworks, Primavera Project Planner, and Bentley STAAD Pro*

Animation and Styling*: Autodesk 3ds Max, Autodesk 3ds Max Design, Autodesk Maya, Autodesk Alias, Foundry NukeX, and MAXON CINEMA 4D*

Computer Programming*: C++, VB.NET, Oracle, AJAX, and Java*

*For more information, please visit the following link: **http://www.cadcim.com***

Note
If you are a faculty member, you can register by clicking on the following link to access the teaching resources: ***http://www.cadcim.com/Registration.aspx***. The student resources are available at ***http://www.cadcim.com***. We also provide **Live Virtual Online Training** on various software packages. For more information, write us at ***sales@cadcim.com***.

Table of Contents

This page is intentionally left blank

Preface

NX 11.0

NX 11.0, a product of SIEMENS Corp., is one of the world's leading CAD/CAM/CAE packages. Being a solid modeling tool, it not only unites 3D parametric features with 2D tools, but also addresses every design-through-manufacturing process. Besides providing an insight into the design content, the package promotes collaboration between companies and provides them an edge over their competitors.

In addition to creating solid models and assemblies, the 2D drawing views can also be generated easily in the **Drafting** environment of NX. The drawing views that can be generated include orthographic, section, auxiliary, isometric, and detail views. The model dimensions and reference dimensions in the drawing views can also be generated. The bidirectionally associative nature of this software ensures that the modifications made in the model are reflected in the drawing views and vice-versa. In NX, you can create sketches directly in the Modeling environment.

The **Mold Design Using NX 11.0 : A Tutorial Approach** textbook has been written with the intention of helping the readers effectively use the mold wizard tools such as gate, runner, and various standard parts in NX. The users will be able to create mold design easily and effectively through some processes such as analysis and documentation which have been dealt in detail. Also, the chapters in this textbook are arranged in a pedagogical sequence that makes this textbook very effective in learning the features and capabilities of the software. The main features of this textbook are as follows:

- **Tutorial Approach**

    The author has adopted the tutorial point-of-view and the learn-by-doing approach throughout the textbook. This approach guides the users through the process of creating the models in the tutorials.

- **Real-World Projects as Tutorials**

    The author has used about 50 real-world mechanical engineering projects as tutorials in this book. This enables the readers to relate the tutorials to the models in the mechanical engineering industry. In addition, there are about 32 exercises that are also based on the real-world mechanical engineering projects.

- **Tips and Notes**

    The additional information related to various topics is provided to the users in the form of tips and notes.

- **Learning Objectives**
  The first page of every chapter summarizes the topics that are covered in that chapter.

- **Self-Evaluation Test, Review Questions, and Exercises**
  Every chapter ends with a Self-Evaluation Test so that the users can assess their knowledge of the chapter. The answers to the Self-Evaluation test are given at the end of the chapter. Also, the Review Questions and Exercises are given at the end of each chapter and they can be used by the Instructors as test questions and exercises.

Formatting Conventions Used in the Textbook

Please refer to the following list for the formatting conventions used in this textbook.

- Names of tools, buttons, options, gallery, and toolbar are written in boldface.

  Example: The **Shrinkage** tool, the **OK** button, the **Main** gallery, and so on.

- Names of dialog boxes, drop-downs, drop-down lists, list boxes, areas, edit boxes, check boxes, and radio buttons are written in boldface.

  Example: The **Scale Body** dialog box, the **Type** drop-down list of **Scale Body** dialog box, the **Uniform** edit box of **Scale Body** dialog box, the **Concept Design** check box in **Standard Part Management** dialog box, the **Add Instance** radio button of the **Standard Part Management** dialog box, and so on.

- Values entered in edit boxes are written in boldface.

  Example: Enter **1.005** in the **Uniform** edit box.

- Names and paths of the files saved are italicized.

  Example: *c03tut03.prt*, *C:\NX_11.0\c03*, and so on.

- The methods of invoking a tool/option from the **Menu**, **Ribbon**, are enclosed in a shaded box.

Ribbon:	Home > Standard > New
Menu:	File > New

Naming Conventions Used in the Textbook

Button

The item in a dialog box that has a 3D shape like a button is also termed as **Button**. For example, **OK** button, **Cancel** button, **Apply** button, and so on. Refer to Figure 1 given next for the terminology used for the components in a dialog box.

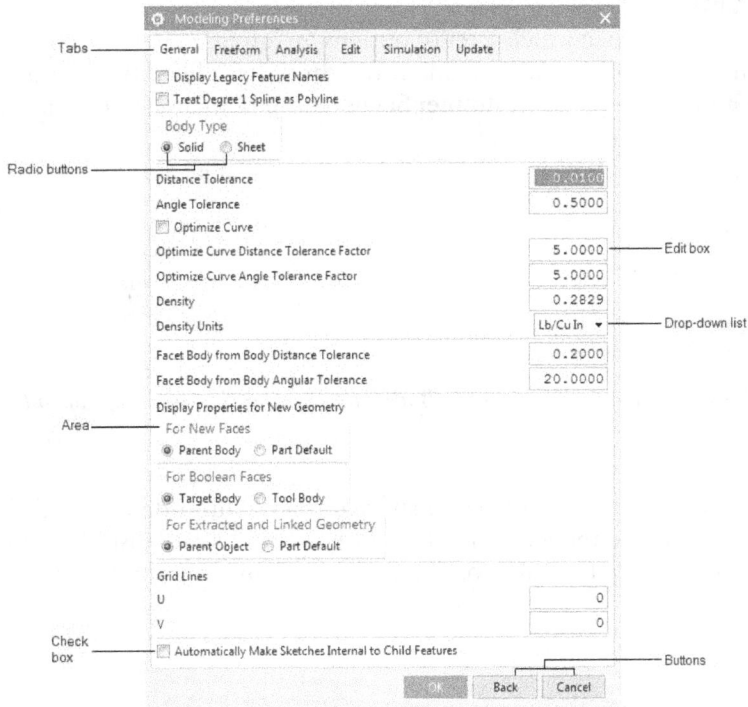

Figure 1 *The components in a dialog box*

Gallery

A gallery is the one in which a set of common tools are grouped together. Refer to Figure 2 for gallery in Mold Wizard.

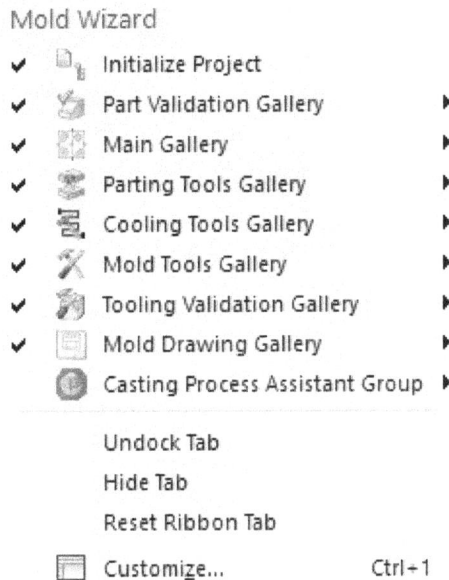

Figure 2 *Gallery in Mold Wizard*

Drop-down List

A drop-down list is the one in which a set of options are grouped together. You can set various parameters using these options. You can identify a drop-down list with a down arrow on it. For example, **Scale Body** drop-down list, **Runner Section** drop-down list, and so on; refer to Figure 3.

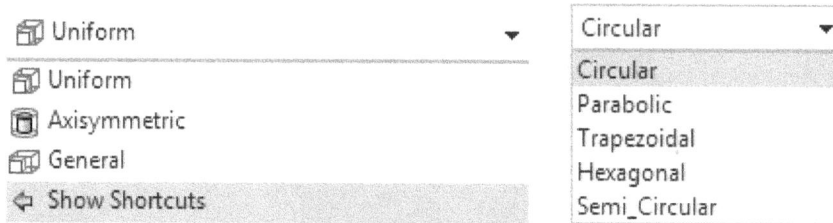

Figure 3 *Partial view of* *Scale Body* *and* *Runner Section* *drop-down lists*

Options

Options are the items that are available in shortcut menu, drop-down list, dialog boxes, and so on. For example, choose the **Fit** option from the shortcut menu displayed on right-clicking in the drawing area; choose the **Faces** option from the **Type** drop-down list; refer to Figure 4.

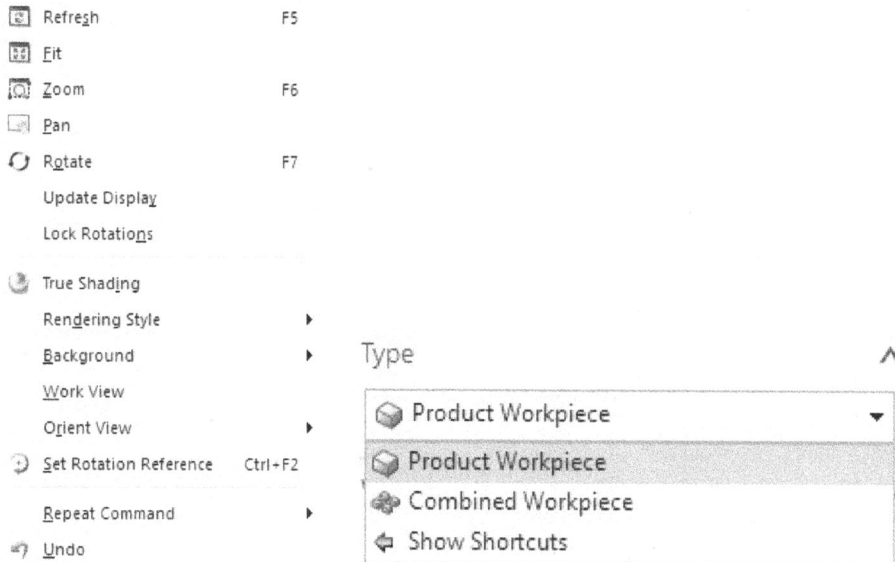

Figure 4 *Options in the shortcut menu and the* *Type* *drop-down list*

Symbols Used in the Textbook

Note
The author has provided additional information related to various topics in the form of notes.

Tip
The author has provided a lot of useful information about the topic being discussed in the form of tips.

Free Companion Website

It has been our constant endeavor to provide you the best textbooks and services at affordable price. In this endeavor, we have come out with a Free Companion website that will facilitate the process of teaching and learning of Mold design using NX 11.0. If you purchase this textbook, you will get access to the files on the Companion website.

The following resources are available for the faculty and students in this website:

Faculty Resources

*   **Technical Support**
    You can get online technical support by contacting *techsupport@cadcim.com*.

*   **Instructor Guide**
    Solutions to all review questions in the textbook are provided in this guide to help the faculty members test the skills of the students.

*   **Part Files**
    The part files used in illustrations, tutorials, and exercises are available for free downloading.

Student Resources

*   **Technical Support**
    You can get online technical support by contacting *techsupport@cadcim.com*.

*   **Part Files**
    The part files used in illustrations and tutorials are available for free downloading.

If you face any problem in accessing these files, please contact the publisher at *sales@cadcim.com* or the author at *stickoo@pnw.edu* or *tickoo525@gmail.com*.

Stay Connected
You can now stay connected with us through Facebook and Twitter to get the latest information about our textbooks, videos, and teaching/learning resources. To stay informed of such updates, follow us on Facebook (*www.facebook.com/cadcim*) and Twitter (*@cadcimtech*). You can also subscribe to our YouTube channel (*www.youtube.com/cadcimtech*) to get the information about our latest video tutorials.

This page is intentionally left blank

Chapter 1

Introduction to Mold Design and NX Mold Wizard

Learning Objectives

After completing this chapter, you will be able to:

- *Understand injection molding*
- *Understand types of mold*
- *Understand different parts of injection mold*
- *Understand injection molding machine*
- *Understand types of injection molding machine*
- *Understand prerequisites to NX Mold Wizard*
- *Understand important terms in NX Mold Wizard*

INTRODUCTION

NX Mold Wizard is an application which provides tools to design plastic injection molds. In this book, You will learn the design process of Injection Mold is discussed.

NX 11.0 Mold Wizard, a product of SIEMENS Corp., provides advanced modeling tools for creating the components of mold such as cores, cavities, parting surfaces, cooling channels, sliders, lifters, and sub inserts. Due to these advanced modeling tools, it is easy to use and design the 3D associative injection mold and also the changes made in the component are automatically updated in the mold.

Mold Wizard contains a mold base library and a standard parts library which help in placing parts associatively. You can customize the standard parts library as per your requirement and can use the library to shorten the tool design and also the manufacturing cycle time. Also, you can accelerate and optimize the mold design process by using standardized processes and components. NX Mold Wizard streamlines the development of mold design process from part design to tool assembly layout, tool design, and tool validation.

Using NX Mold Wizard, you can analyse a part for its moldability and manufacturability. You can also check mold assembly for clearance and relief in various position states. To understand the mold design, you should be familiar with some basic concepts such as the injection molding machine, part defects, manufacturing processes, gate and runner types, types of ejection, and some basic calculations, which are discussed in the later chapters.

BASIC CONCEPT OF MOLD DESIGN

Injection mold is an assembly of parts containing an impression into which plastic material is injected and cooled to create a component. You can process thermoplastic as well as thermosetting polymers to create parts in injection molding machine. Injection mold enables in producing large numbers of parts of high quality very quickly with accuracy and efficiency.

To ensure the smooth workflow of the mold design process, the injection mold parts must be designed carefully. The material used for the part and the mold, the desired shape and features of the part, and the properties of the molding machine must all be taken into account before mold design.

Types of Injection Mold

The DIN IS0 standard 12165, "Components for Compression, Injection, and Compression-Injection Molds" classifies molds as follows:

1. Two Plate Mold, refer to Figure 1-1
2. Three Plate Mold, refer to Figure 1-2
3. Stripper Plate Mold, refer to Figure 1-3
4. Split-Cavity Mold, refer to Figure 1-4
5. Stack Mold, refer to Figure 1-5
6. Hot Runner Mold, refer to Figure 1-6

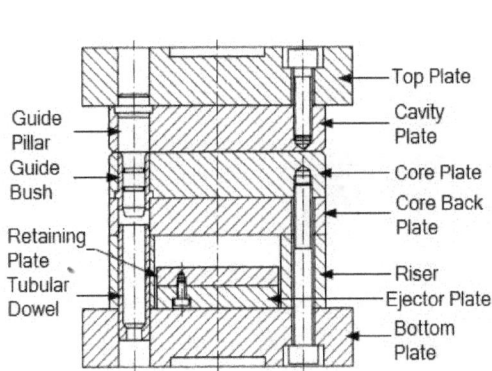

Figure 1-1 *The structure of a two plate mold*

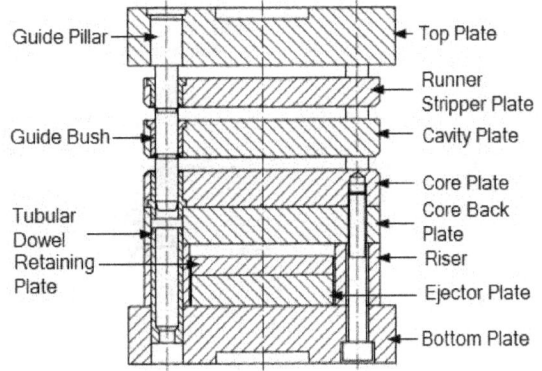

Figure 1-2 *The structure of a three plate mold*

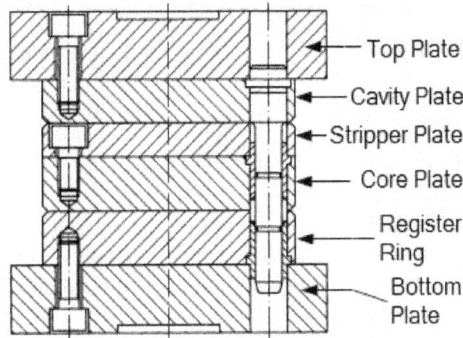

Figure 1-3 *The structure of stripper plate mold*

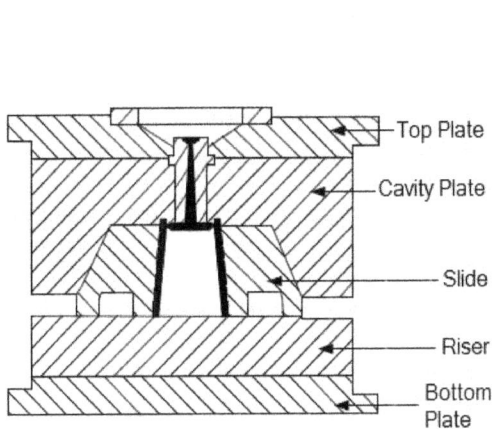

Figure 1-4 *The structure of split-cavity mold*

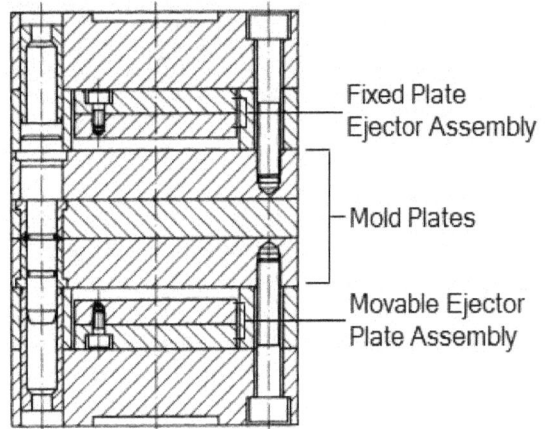

Figure 1-5 *The structure of stack mold*

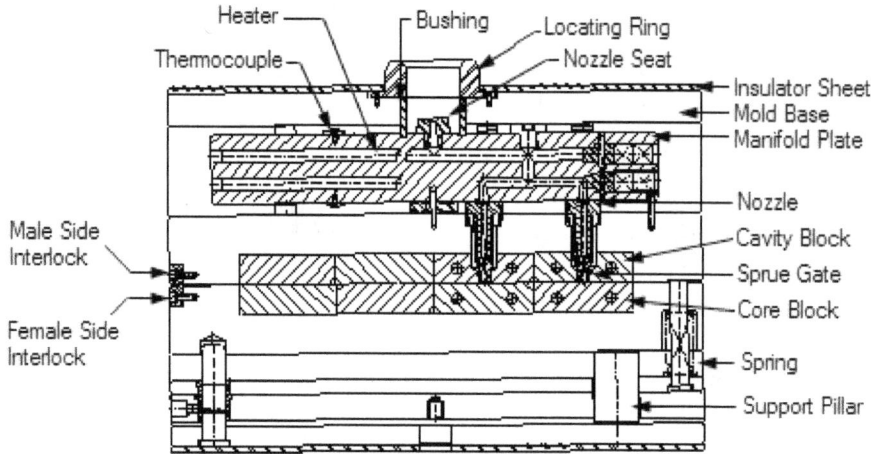

Figure 1-6 *The structure of hot runner mold*

Parts of Injection Mold

Injection mold consists of various parts, refer to Figure 1-7. Some of the parts are discussed next.

- **Register Ring**: Register ring is a circular member which is fitted into the front face of the mold plate and is used to locate the mold in correct position on the injection machine. This also ensures proper alignment between Nozzle and Sprue bush to prevent leakage of material.

- **Sprue Bush**: Sprue bush is a connecting member between machine nozzle and mold face which has a suitable aperture through which the molten material travels into the cavity.

- **Top Plate**: Top plate is used to clamp the mold with the injection molding machine on a stationary platen.

- **Bottom Plate**: Bottom plate is used to clamp the mold with the injection molding machine on a movable platen.

- **Core Plate**: It is a plate or block of steel which contains the core. The core may be machined directly from a solid plate or may be a two-part construction.

- **Cavity Plate**: It is a plate or block of steel which contains the cavity. The cavity may be machined directly from a solid plate or may be a two-part construction.

- **Ejector Plate**: It is a plate which is used for ejecting the component.

- **Ejector Retention Plate**: It is a plate which contains ejector pin for ejecting the component.

Sprue Bush — Register Ring
Cooling Pipe → Top Plate
Wear Plate
Slider Cavity Plate
Angle Pin Guide Pillar
Sprue Puller Pin Core Plate
Guide Bush
Support Pillar Core Back Plate

Ejector Retention Plate
Ejector Plate

Ejector Rod Bush
Bottom Plate

Ejector Rod

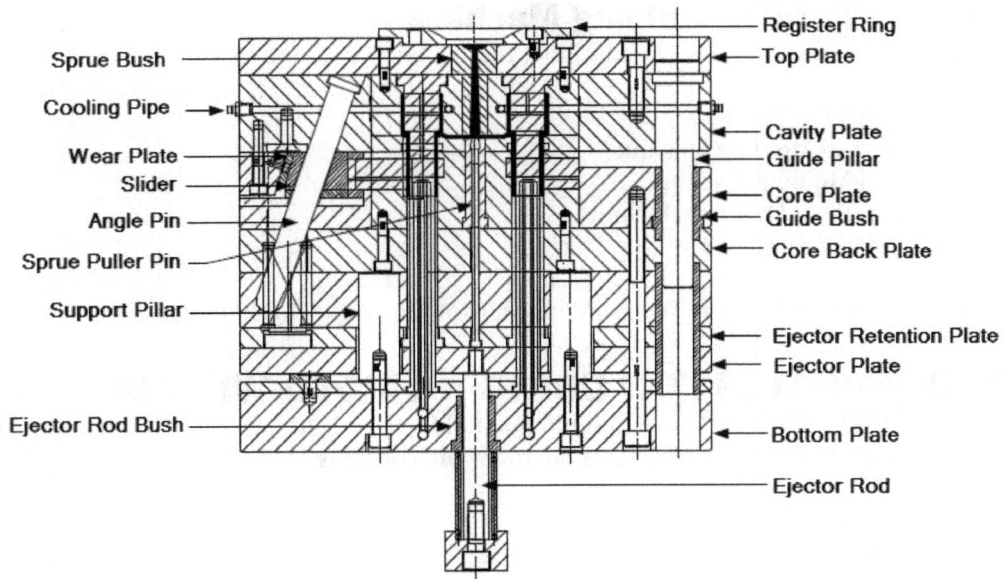

Figure 1-7 Different parts of Injection Mold

Injection Molding Machine

Injection molding machine is used to melt the molding plastic material inside the heating cylinder and then injecting this material into mold to create molded product by solidifying the material. Various parts of Injection molding machine are shown in Figure 1-8.

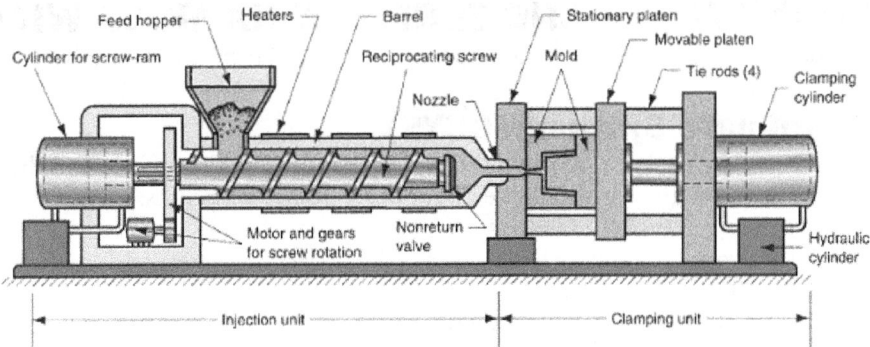

Feed hopper   Heaters   Barrel   Stationary platen
Cylinder for screw-ram   Reciprocating screw   Mold   Movable platen
Nozzle   Tie rods (4)   Clamping cylinder

Motor and gears   Nonreturn   Hydraulic
for screw rotation   valve   cylinder

Injection unit   Clamping unit

Figure 1-8 Different parts of Injection Molding Machine

Types of Injection Molding Machine

According to the orientation of the machine, the Injection molding machines are classified as follows:

(a) Horizontal Molding Machine
(b) Vertical Molding Machine
(c) Rotary Table Molding Machine
(d) Two Stage Injection Molding Machine

In this book, you will design injection molds only for Horizontal Molding Machine.

PREREQUISITES FOR USING NX MOLD WIZARD

Prior to using NX Mold Wizard for designing a mold, you should have a hands on experience on NX and must have understanding of the following concepts and processes:

- Feature Modeling
- Free Form Modeling
- Curves
- Layers
- Assemblies and Assembly Navigator
- Changing the Display and Work Part
- Adding and Creating Components
- Creating and Replacing Reference Sets
- Drafting

IMPORTANT TOOLS AND TERMS IN NX MOLD WIZARD

In this section, you will learn about some important tools and terms used in NX Mold.

Mold Coordinate System(MCSYS)

The **Mold CSYS** tool helps you to reorient the part with respect to the mold assembly. To define the mold coordinate system, you should be aware of the following points:

- Orient the product model such that the ejection direction corresponds to the Z - axis of the mold base.
- Position the product model such that the principal parting plane lies on the XY-plane of the mold base.
- Position the product model on the XY-plane of the mold base.

Shrinkage

Shrinkage is a process where the molten plastic filled inside the cavity of a mold shrinks at the time of solidification. The **Shrinkage** tool helps you to apply the scale factor on the product model. To define the scale factor, you should be aware of the following points:

- Type of Molding material
- Wall thickness of the molded part
- Cavity surface temperature

- Gate shape
- Presence of additive material in the molding materials

Workpiece

This tool helps you define the insert size of core and cavity. As the size of the insert increases the cost of the mold also increases because the material of the insert is costlier as compared to the cost of the plate. So, you need to be careful while deciding the size of the insert. The material generally used is P20 which has a good toughness and hardness level.

Cavity Layout

This tool and its layout type options help you to design multi-cavity mold. Its layout type can affect the mold filling time, so you need to be careful when you select the layout type.

Gate

Gate is a channel which connects the runner with the impression. The size and type of gate influence the cycle time of injection molding. The positioning of gate should be such that there is an even flow of molten material in the impression and it fills uniformly and reaches all impression extremities at the same time.

The **Gate Library** tool helps you to position the gate, change the size, and select the type of gate. The positioning settings determines the orientation of the gate. The gates are fully associative with their pockets.

The types of gates available in Mold Wizard are fan, film, pin, pin point, rectangle, step pin, tunnel, and curved tunnel, refer to Figure 1-9 (a) to (h).

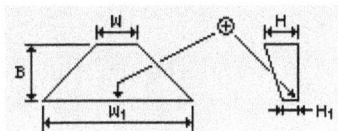

fan gate

(a)

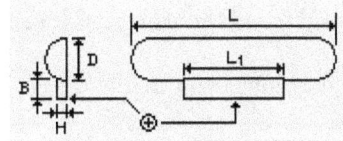

film gate

(b)

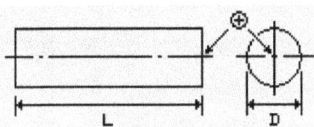

pin gate

(c)

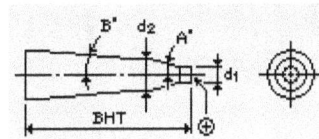

pin point

(d)

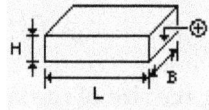

rectangle gate

(e)

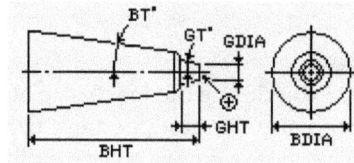

step pin

(f)

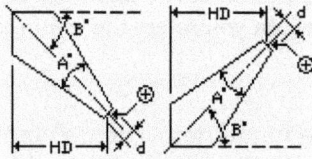

core side   cavity side

tunnel gate

(g)

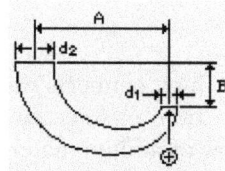

curved tunnel

(h)

Figure 1-9 Various types of gates

Runner

Runner is a channel that is machined into the mold plate to connect the sprue with the gate to the impression. While designing the runner, you should keep in mind the shape of the cross- section, size of the runner, and the runner layout.

The **Runner** tool helps you define the section type, parameters of section, and the layout.

There are various types of runners available in Mold Wizard such as Circular, Parabolic, Trapezoidal, Hexagonal, and Semi Circular, as shown in Figure 1-10 (a) to (e).

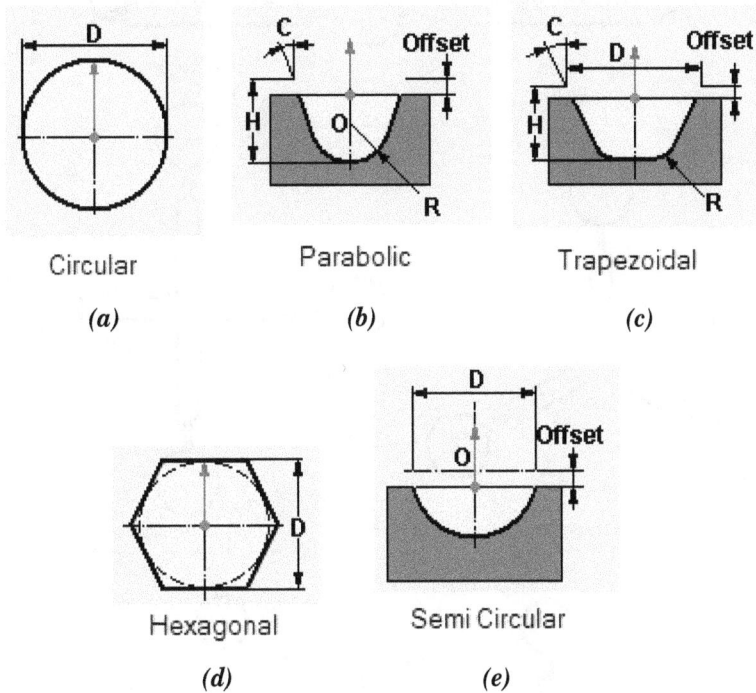

Figure 1-10 *Various types of runners*

Runner Layout

The layout of a runner system depends on the number of impressions, shape of the components, type of mold, and type of gate.

While designing the runner layout, keep length of the runner minimum to reduce pressure losses. Runner system should be balanced. Figure 1-11 shows examples of molds based on the balanced runner principle.

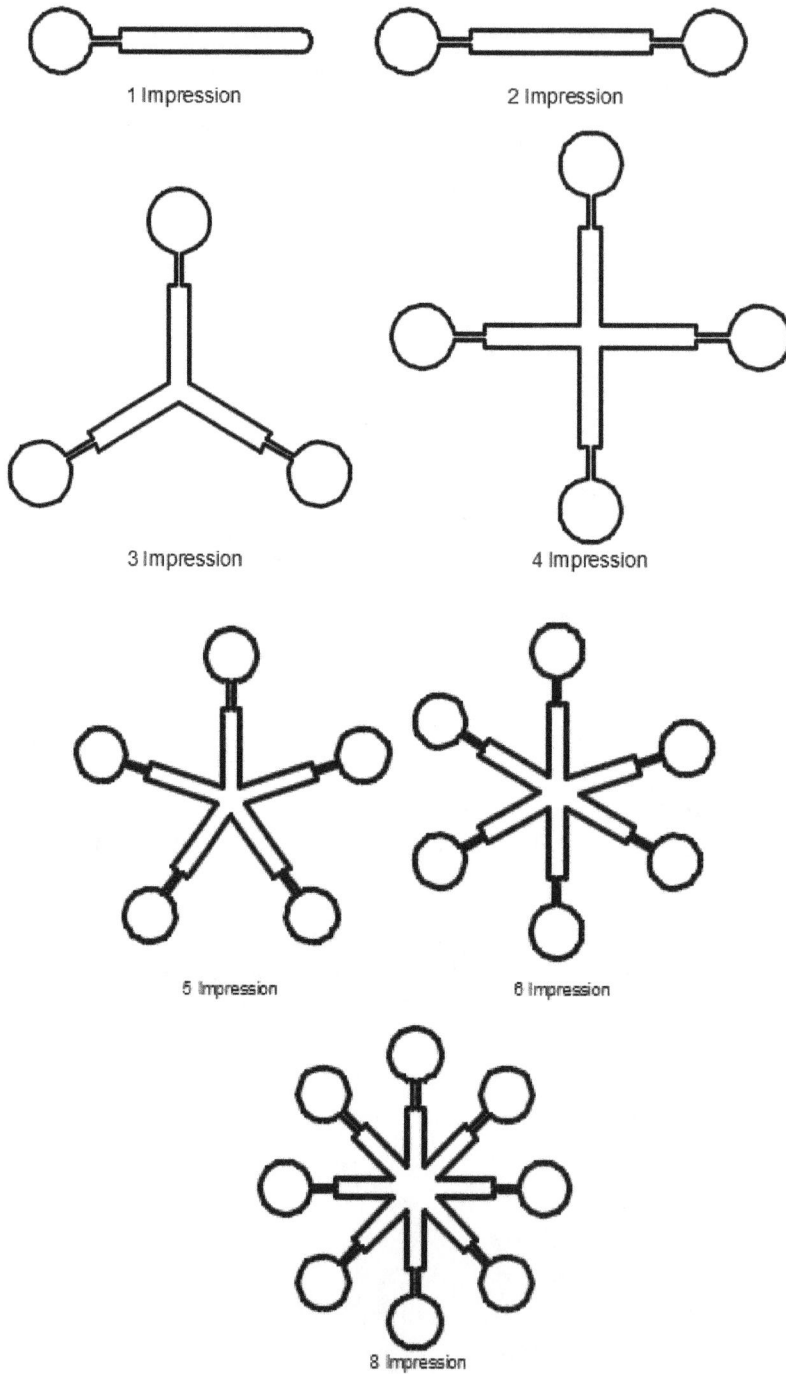

Figure 1-11 The balanced runner layouts

Cooling

Cooling is a process used for solidifying the material. The location of cooling line depends on the part geometry, cavity configuration, location of ejector pins, and moving components of the mold.

To use cooling channel for cavity cooling and core cooling, there are two different methods. Figure 1-12 shows how to layout cooling channels in cavity and Figure 1-13 shows how to layout cooling channels in core. There is one inlet and one outlet for a coolant and the arrows in the figures show the direction of coolant flow. Notice that to create a cooling channel for the core, you need to drill on plate at an angle which is quite difficult.

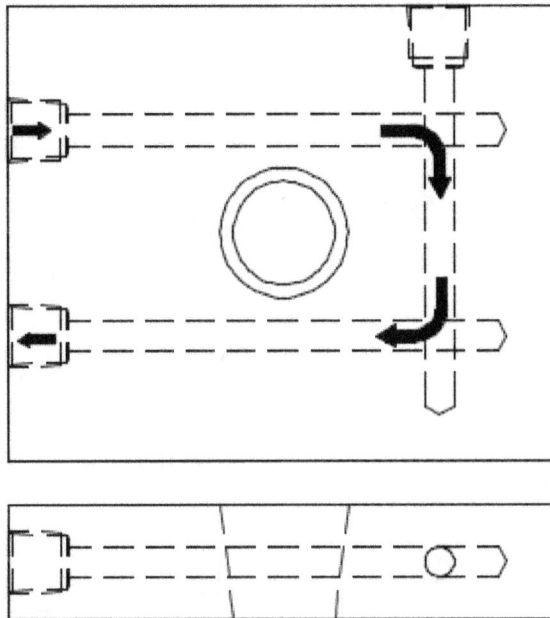

Figure 1-12 Cooling circuit design for cavity

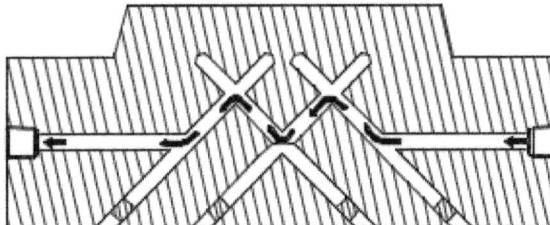

Figure 1-13 Cooling circuit design for core

In the Mold wizard, all the tools related to cooling are available in the **Cooling Tools** gallery. In this wizard, you will create channels, adjust channels, and add fittings to the channels.

Ejection

Ejection is a process that is used for removing the component from the mold. To eject a component, you need to add ejector pins.

There are various types of pins which you can select for component ejection according to the type of ejection required. But, in general, the choice depends upon the shape of the molding part. Some ejection techniques are as follows:

(i)   Pin Ejection
(ii) Sleeve Ejection
(iii) Bar Ejection
(iv) Blade Ejection
(v)   Air Ejection
(vi) Stripper Plate Ejection

Figure 1-14 shows various types of pin used by Misumi company in Mold Wizard.

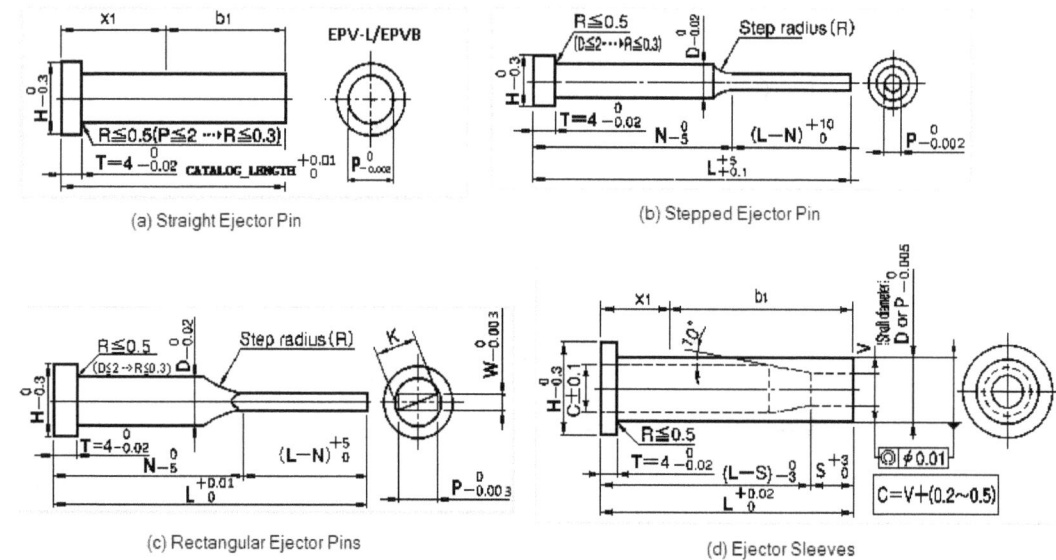

Figure 1-14 Various ejector pins used by MISUMI

Guide Pillar and Guide Bush

Guide pillar and Guide bush are used for guiding the two halves of the mold (fixed half and moving half). There is a sliding fit between guide pillar and guide bush (H7/g6).

To insert the guide pillar and guide bush in the mold, choose the Standard Part Library tool available in the Main Gallery. Figure 1-15 shows various types of guide pillar and guide bush by DME.

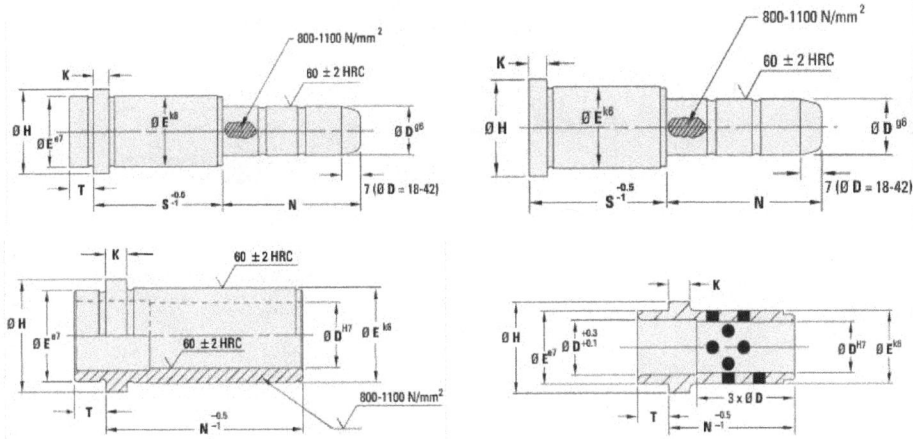

Figure 1-15 *Various guide pillar and guide bush by DME*

GETTING STARTED WITH NX MOLD WIZARD

Before starting NX Mold Wizard, you need to add NX Mold Wizard library to NX. After adding the library to NX, you can use standard parts and material library to create a mold. Start NX by double-clicking on the shortcut icon of NX on the desktop of your computer. After loading all the required files to start NX, the initial interface will be displayed, as shown in Figure 1-16.

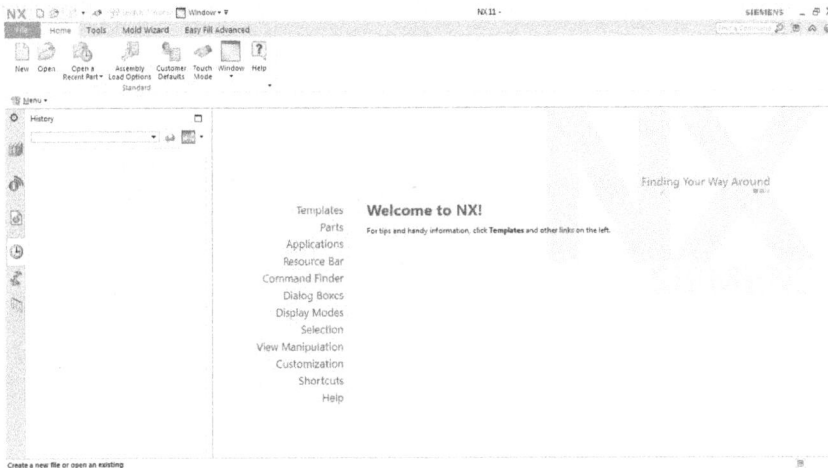

Figure 1-16 *The initial interface that appears after starting NX*

Now, choose **File > Open** from the **Ribbon**; the **Open** dialog box will be displayed, refer to Figure 1-17.

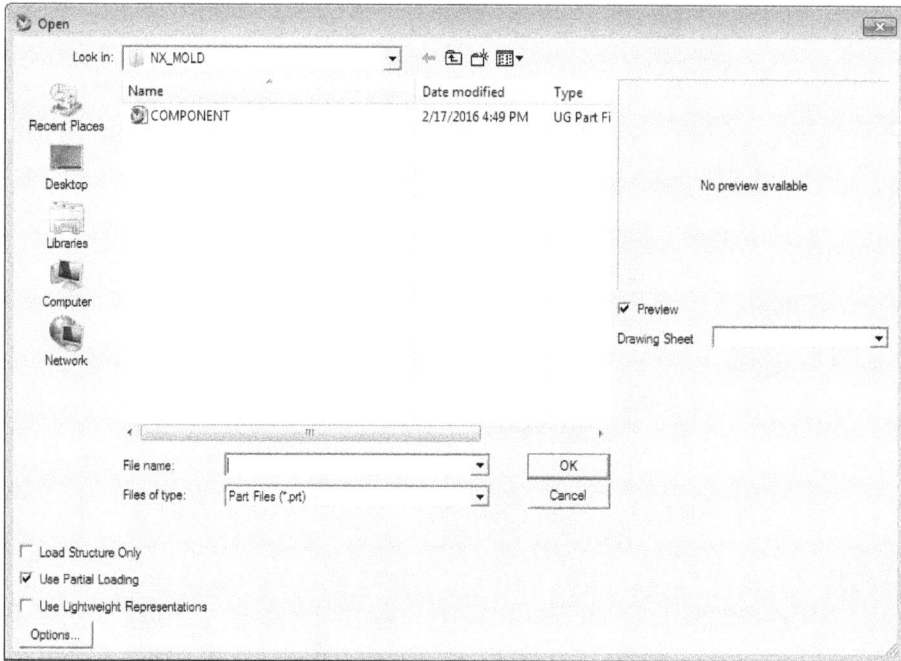

Figure 1-17 The **Open** dialog box

Next, select the file that need to be opened and choose the **OK** button; the file will be opened in the modeling environment, refer to Figure 1-18.

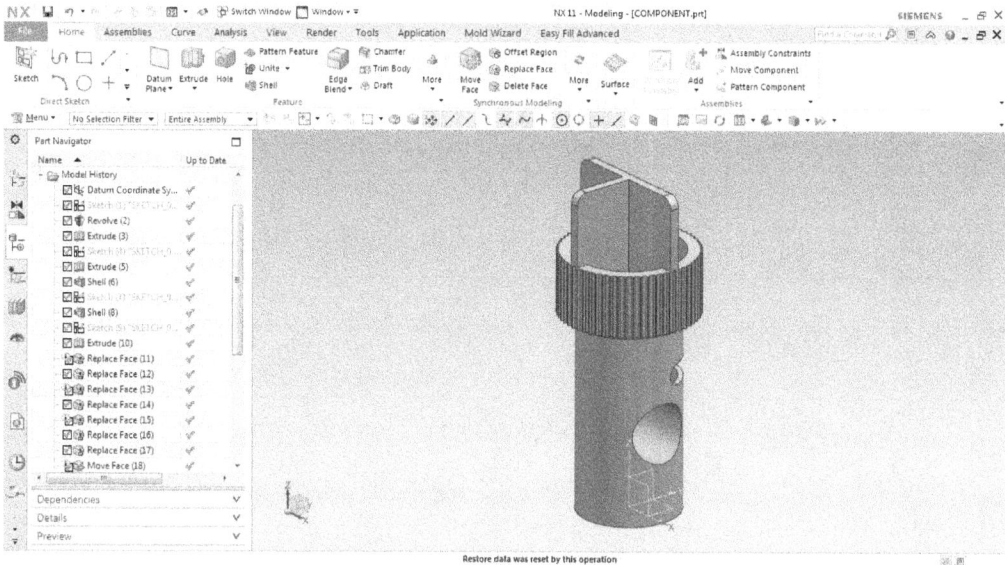

Figure 1-18 A component displayed in the modeling environment

To go to the NX Mold Wizard, choose the **Application** tab and then choose the **Mold** tool from the **Process Specific** group; the NX Mold Wizard environment will be displayed, refer to Figure 1-19.

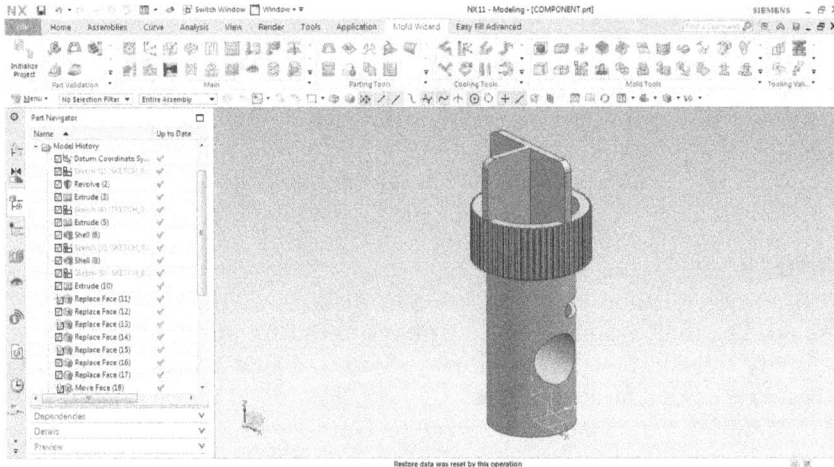

Figure 1-19 A component displayed in the Mold wizard environment

MOLD WIZARD TAB

In NX Mold Wizard, most of the tools are arranged in a sequential manner according to the flow of process. NX Mold offers a user-friendly design interface by providing the galleries and groups. In galleries and groups, various tools and options are grouped together based on their functionality. Various options in the Mold Wizard tab are discussed next.

The Mold Wizard Environment can be invoked from Modeling environment. To do so, choose **Application > Process Specific > Mold** from the **Ribbon**.

Initialize Project

This tool is used to create a new mold project. With the help of this tool, you can specify the type of template assembly, units, path to project part file, project name, material, and shrinkage factor.

Part Validation Gallery

This gallery comprises of validation tools which are used to check the quality of the molded parts. The tools used in this gallery are **Mold Design Validation**, **Check Regions**, **Check Wall Thickness**, **Run Flow Analysis**, and **Display Flow Analysis Results**, refer to Figure 1-20.

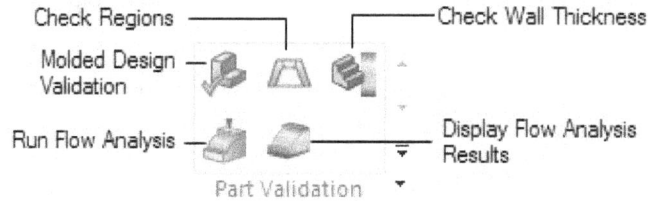

*Figure 1-20 The **Part Validation** gallery*

Main Gallery

This gallery comprises of tools which are used to make gate, runners, sliders, lifters, mold plates, and so on. The tools available in this gallery are **Family Mold**, **Mold CSYS**, **Shrinkage**, **Workpiece**, **Cavity Layout**, **Mold Base Library**, **Standard Part Library**, **Design Ejector Pin**, **Ejector Pin Post Processing**, **Slide and Lifter Library**, **Sub-insert Library**, **Design Fill**, **Runner**, **Pocket**, **Bill of Material**, and so on, refer to Figure 1-21.

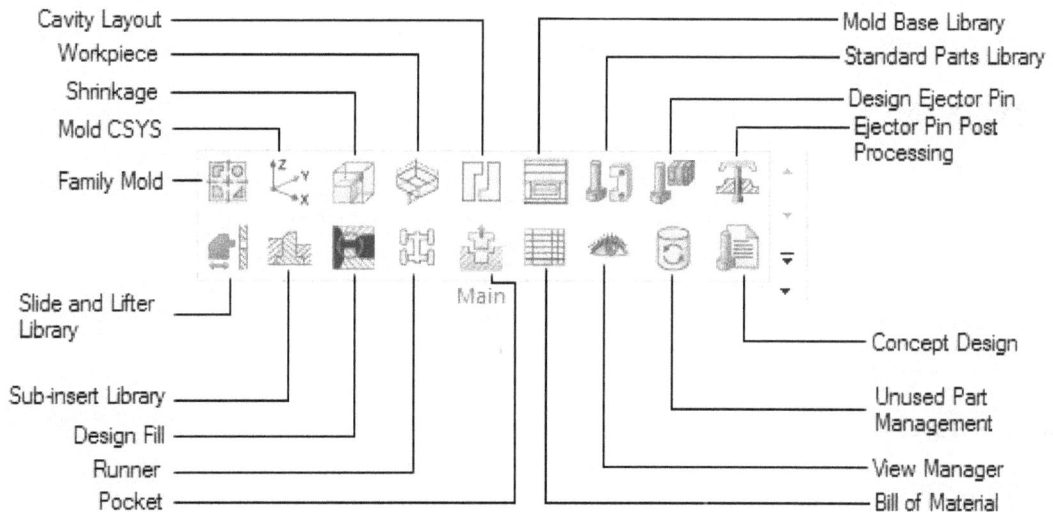

*Figure 1-21 The **Main** gallery*

Parting Tools Gallery

This gallery consists of tools which are used to create parting surface, core and cavity of the part. The tools in this gallery are **Check Regions, Patch Surface, Define Regions, Design Parting Surface, Edit Parting and Patch Surface, Define Cavity and Core, Swap Model, Back Up Parting/Patch Sheets**, and **Parting Navigator**, refer to Figure 1-22.

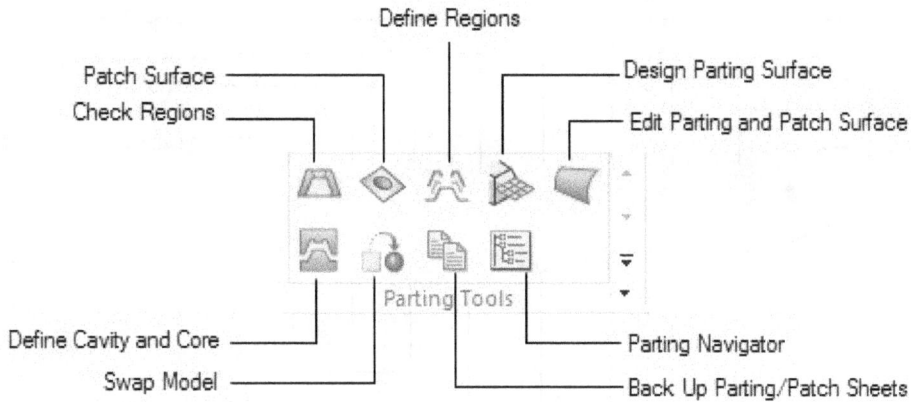

*Figure 1-22 The **Parting Tools** gallery*

Cooling Tools Gallery

This gallery consists of tools which are used to create cooling channels for cooling the material. The tools in this gallery are **Pattern Channel**, **Direct Channel**, **Define Channel**, **Connect Channels**, **Extend Channel**, **Adjust Channel**, **Cooling Fittings**, **Cooling Circuits**, and **Cooling Standard Part Library**, refer to Figure 1-23.

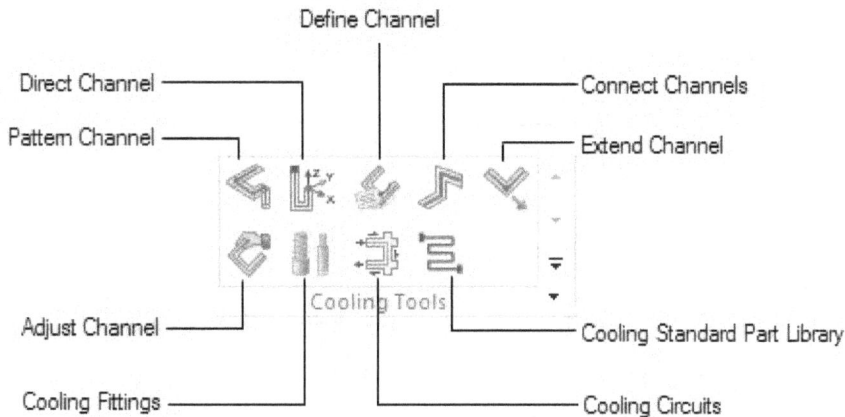

*Figure 1-23 The **Cooling Tools** gallery*

Mold Tools Gallery

This gallery consists of tools which are used to make mold design parts faster. The tools in this gallery are **Bounding Body**, **Split Body**, **Solid Patch**, **Trim Region Patch**, **Enlarge Surface Patch**, **Guided Extension**, **Extend Sheet**, **Split Face**, and so on, refer to Figure 1-24.

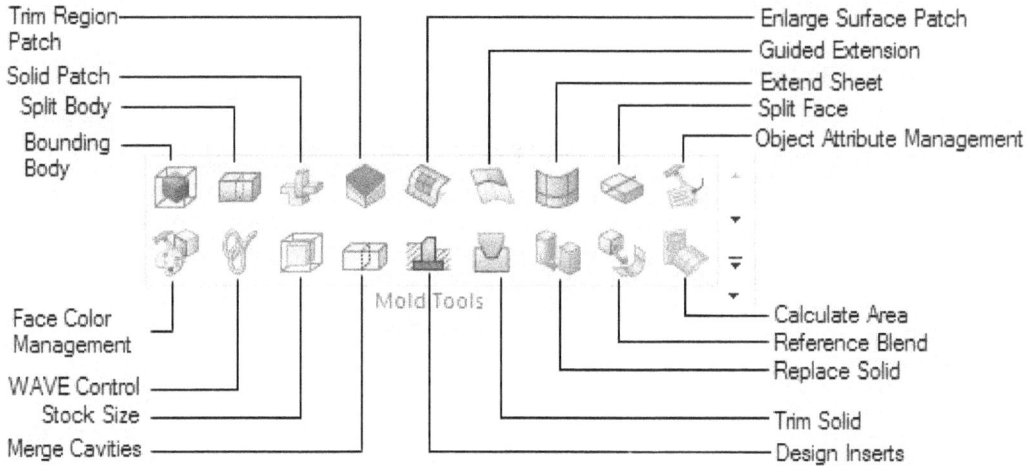

Figure 1-24   *The **Mold Tools** gallery*

Tooling Validation Gallery

This gallery consists of tools which are used for applying simulation to the mold assembly. The tools in this gallery are **Static Interference Check, Preprocess Motion, Define Slide, Define Lifter, User Defined Motion, Run Simulation,** and so on, refer to Figure 1-25.

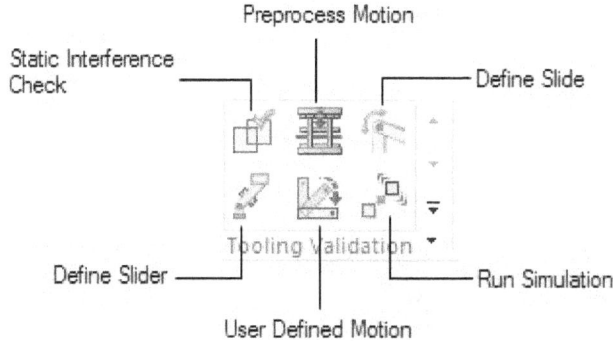

Figure 1-25   *The **Tooling Validation** gallery*

Mold Drawing Gallery

This gallery consists of tools which are used for creating detailed drawing of a mold. The tools in this gallery are **Assembly Drawing, Component Drawing, Hole Table, Auto Dimension, Hole Manufacturing Note, Ejector Pin Table** and so on, refer to Figure 1-26.

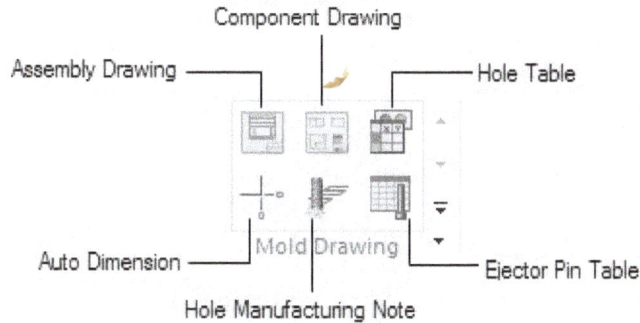

*Figure 1-26 The **Mold Drawing** gallery*

Casting Process Assistant Group (Customize to add)

This group consists of tools which are used to model product and tooling features for molded parts, refer to Figure 1-27.

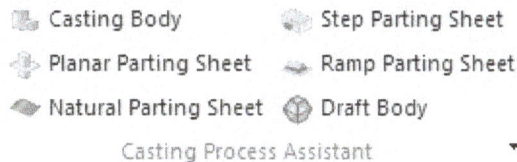

*Figure 1-27 The **Casting Process Assistant** group*

Easy Fill Advanced

To analyze the part, you need to add the **Easy Fill Advanced** plugin to NX. It helps you to do various things such as multi-gate analysis, packing, shrinkage, and fiber orientation, which are needed to validate mold design prior to manufacturing. These are discussed in detail in the next chapter.

Self-Evaluation Test

Answer the following questions and then compare them to those given at the end of this chapter:

1. You can process thermoplastic and _____ polymers to create component in injection molding machine.

2. There are _____ types of injection molds according to DIN IS0 standard 12165.

3. Register ring is _____ in shape.

4. There are _____ types of injection molding machines.

5. The _____ gallery is used to check the molded part quality.

6. The _____ gallery is used to create cooling channels.

7. Shrinkage is used to apply scale factor to the components. (T/F)

8. Gate is a channel which connects sprue bush to the components. (T/F)

9. Runner is a channel which connects sprue bush to the gate. (T/F)

10. Guide pillar and guide bush have tight fit between them. (T/F)

Review Questions

Answer the following questions:

1. Which of the following members is used to register a mold in the injection molding machine?

    (a) **Register Ring**   (b) **Sprue Bush**
    (c) **Top Plate**   (d) None of these

2. Which of the following members is used to clamp a mold on the stationary side of the injection molding machine?

    (a) Bottom Plate   (b) Top Plate
    (c) Register Ring   (d) None of these

3. Which of the following is not a type of gate according to the mold wizard library?

    (a) **Fan**   (b) **Film**
    (c) **Tunnel**   (d) **Hexagonal**

4. Which of the following is a type of runner according to the mold wizard library?

    (a) **Parabolic**   (b) **Pin**
    (c) **Pin point**   (d) **Step pin**

5. Which type of fit is used between guide pillar and guide bush?

    (a) **H7/s6**   (b) **H7/u6**
    (c) **H7/r6**   (d) **H7/g6**

EXERCISE

Exercise 1

Identify the name of the parts marked with numbers in Figure 1-28.

(Expected time: 10 min)

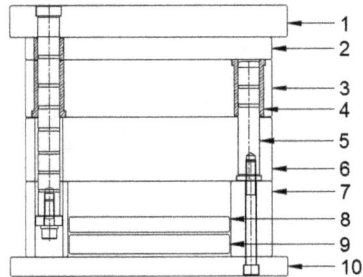

Figure 1-28 *Mold structure*

Answers to Self-Evaluation Test

1. thermosetting, **2.** 6, **3.** circular, **4.** 4, **5. Part Validation**, **6. Cooling Tools**, **7.** T, **8.** F, **9.** T, **10.** F

Chapter 2

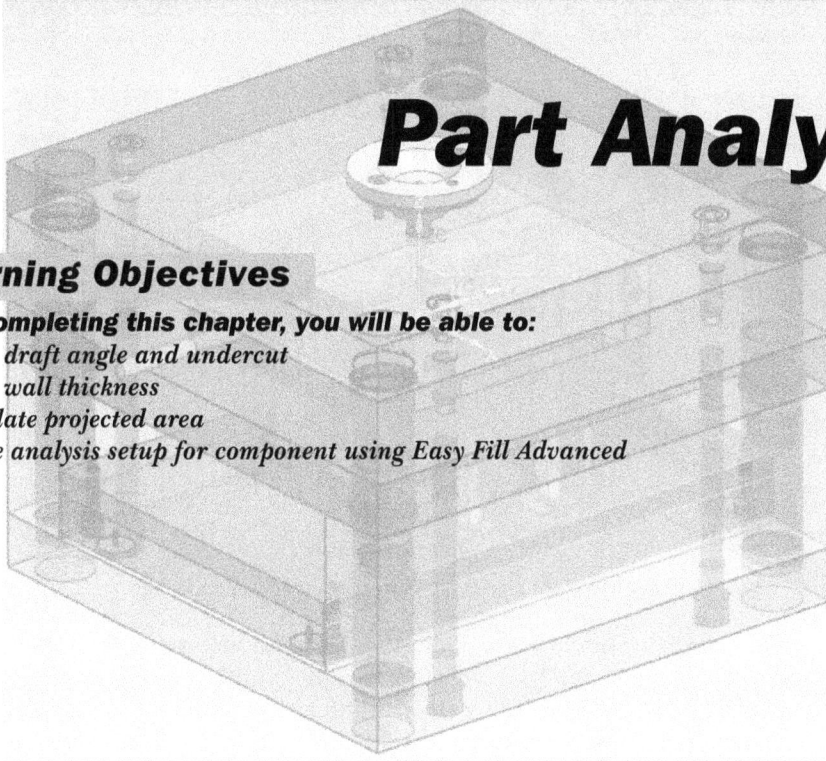

Part Analysis

Learning Objectives

After completing this chapter, you will be able to:

- *Check draft angle and undercut*
- *Check wall thickness*
- *Calculate projected area*
- *Create analysis setup for component using Easy Fill Advanced*

INTRODUCTION

In this chapter, you will learn part analysis which is the first step in mold designing. You will also install the Easy Fill Advanced plugin to analyze any given part.

PART ANALYSIS

Part analysis is the first step of a mold designing process. By analyzing a part, you can better understand the moldability of that part. To analyze a part, you need to install the **Easy Fill Advanced** plugin. As you install the plugin; the **Easy Fill Advanced** tab becomes available in NX interface, refer to Figure 2-1. Now, you can use the tools available in the **Easy Fill Advanced** tab to analyze the part.

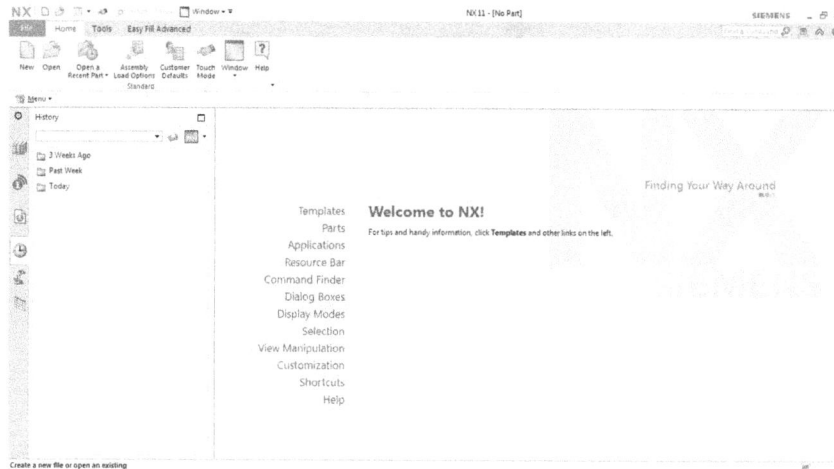

*Figure 2-1 The **Easy Fill Advanced** tab*

Before analyzing the part, you should know about the tools that are available in the NX mold wizard environment. You can use the tools according to the type of analysis that you need to perform. The types of analysis which you will learn are draft analysis, undercut analysis, thickness analysis, projected area, melt front time, air trap, weld line, volumetric shrinkage, shear stress, and cooling time.

Mold Design Validation

Ribbon: Mold Wizard > Part Validation gallery > Mold Design Validation

This tool helps to check the quality of the molded part. To check the quality of undercut, draft angle, and mold part, invoke the **Mold Design Validation** tool from the **Part Validation** gallery of the **Mold Wizard** tab; the **Mold Design Validation** dialog box will be displayed, as shown in Figure 2-2. The options in this dialog box are discussed next.

Component Validation

The check boxes available in this area help you to find out interference between the electrodes and the overlapping of core and cavity sheets.

Product Quality
The check boxes available in this area help you to find out the undercut and the draft angle of the molded part.

Parting Validation
The check boxes available in this area help you to find out the faces to be split and also the overlapping patch surfaces.

As you choose the **Execute Check-Mate** button from the **Parameters** rollout of the dialog box, the **HD3D Tools** window is displayed, as shown in Figure 2-3. Depending upon the check boxes selected in the **Mold Design Validation** dialog box, the parameters will be displayed in the **HD3D Tools** window. In this window, the status of the results is shown in symbols. Following are the symbols with their meanings:

Passed with information.

Passed with warning.

Passed.

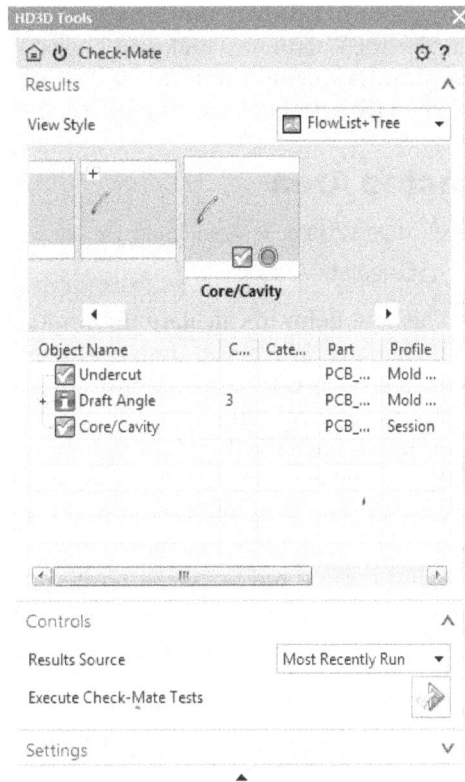

Figure 2-2 The *Mold Design Validation* dialog box

Figure 2-3 The *HD3D Tools* window

Check Wall Thickness

A part with uniform thickness allows the mold cavity to fill more easily as here the molten plastic does not have to be forced through varying restrictions while filling. If the walls of the part are not uniform then the material on thin section cools first than on the thick section which leaves warpage or sink marks in the part. So, the part thickness should ideally be uniform.

You can analyze the thickness using the **Check Wall Thickness** tool. To analyze thickness of the part, invoke the **Check Wall Thickness** tool from the **Part Validation** gallery of the **Mold Wizard** tab; the **Check Wall Thickness** dialog box will be displayed, refer to Figure 2-4, and you will be prompted to select solid body to analyze. Select the solid body and then select the **Ray** or **Rolling Ball** radio button from the **Method** rollout to calculate the thickness of the part. Now, choose the **Calculate Thickness** button from the **Process Results** rollout; the result will be displayed. Choose the **OK** button to close the dialog box.

Projected Area

This tool helps to calculate the projected area of the part. When you design any mold, it is necessary to calculate the optimum required mold clamping force that the injection molding machine should have for the mold to be installed on the machine. For example, if a mold to be installed requires a clamping force of 100 ton and the mold is installed in 75 ton molding machine then the molded product thus created will be full of flash. Similarly, if you install it in 300 ton molding machine, the molding operation would still be possible but the cost of operation will become high.

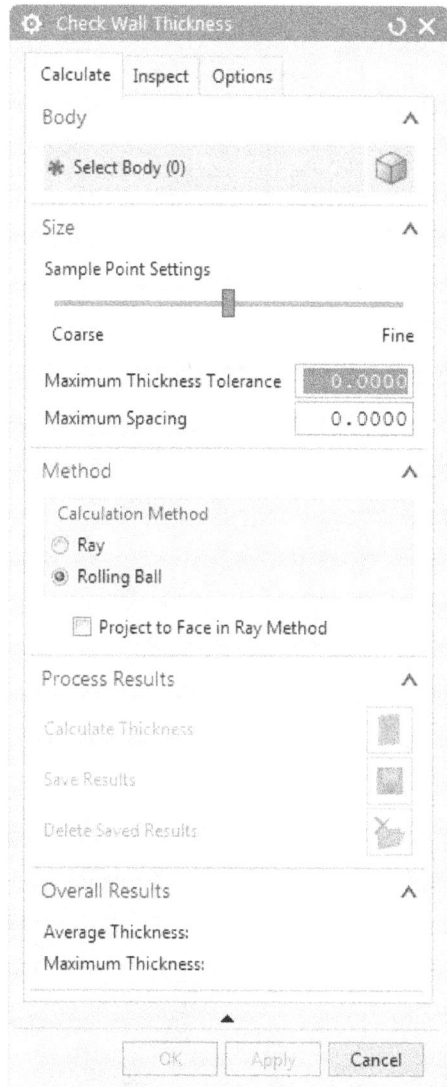

*Figure 2-4 The **Check Wall Thickness** dialog box*

Clamping Force = Projected Area x Injection Pressure

Now, to calculate the projected area of the part, invoke the **Calculate Area** tool from the **Mold Tools** gallery of the **Mold Wizard** tab; the **Calculate Area** dialog box will be displayed, as shown in Figure 2-5, and you will be prompted to select sheet or solid bodies to calculate area. Select the body and then select the **Specify Plane** area; you will be prompted to select objects to define the plane. Select the object face and specify the distance of the plane in the **Distance** edit box. Choose the **OK** button to close the dialog box; the **Information** window will be displayed where projected area calculations are shown, refer to Figure 2-6. Choose the **X** button in the **Information** window to close the window.

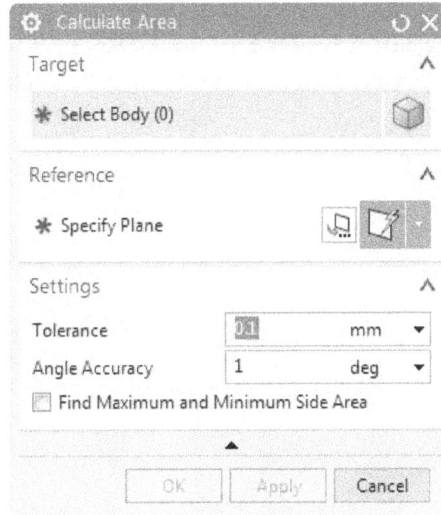

Figure 2-5 The **Calculate Area** *dialog box*

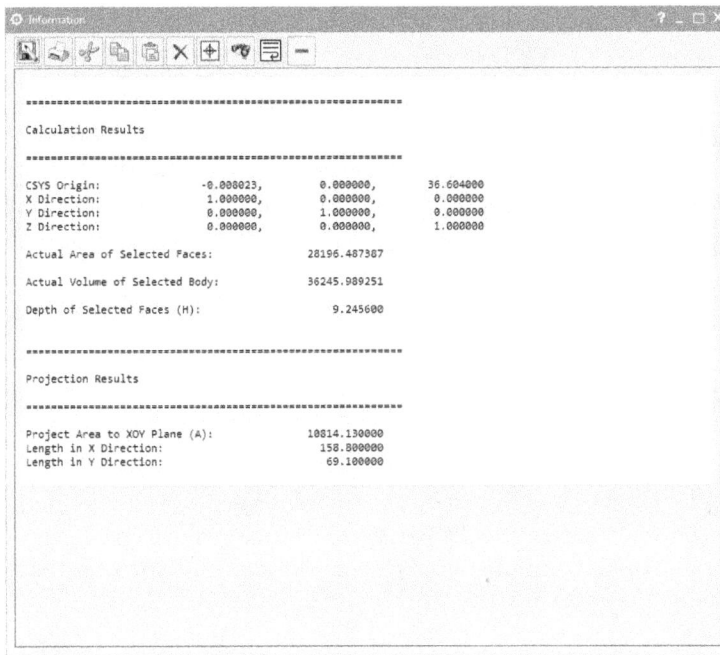

Figure 2-6 The **Information** *window*

For EDM machining, you can use the projected area to estimate the required ram force and the height to calculate the travel distance of ram.

Easy Fill Advanced

NX provides various tools to simulate the flow of material into the mold. To simulate the flow of material, you need to add plugins. Various plugins that are available for simulation are Moldex3D, Easy Fill, and Easy Fill Advanced. In this chapter, you will use Easy Fill Advanced plugin because this provides more advanced options to analyze the part for mold designing. In the tutorial of this chapter, you will use tools available in the Easy Fill Advanced plugin. As you add this plugin; the **Easy Fill Advanced** tab gets displayed in the NX Welcome window, refer to Figure 2-7.

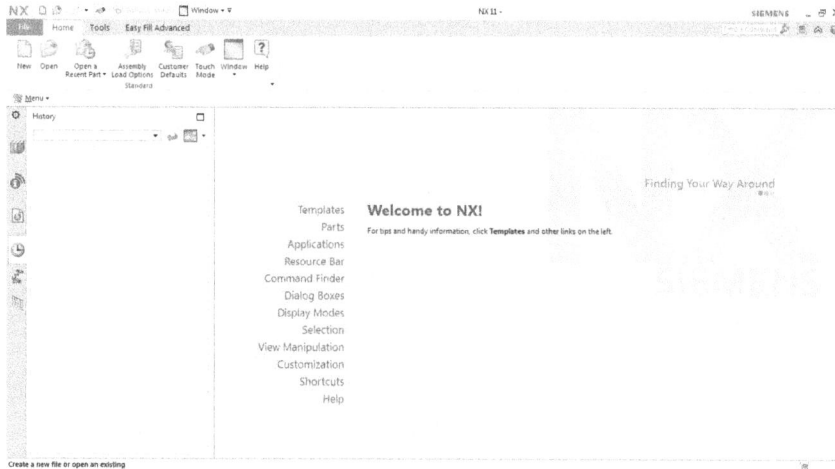

Figure 2-7 *The NX Welcome window with the **Easy Fill Advanced** tab*

TUTORIALS

To perform the tutorials, you need to download the zipped file named as *c02_NX_Mold_input* from the **Input Files** section of the CADCIM website. The complete path for downloading the file is:

> *Textbooks > CAD/CAM > NX_Mold > Mold Design Using NX 11.0: A Tutorial Approach > Input Files*

After the file is downloaded, extract the folder and rename it as *c02*.

Tutorial 1

In this tutorial, you will analyze the PCB_COVER model which you have downloaded and is shown in Figure 2-8. You will also apply gate, runner, and sprue to analyze the flow of material into the model. Next, you will display and analyze the results.

After analysis, save the file with the name *c02tut1.prt* at the location given below:
  \c02   (**Expected time: 3 hr**)

The following steps are required to complete this tutorial:

a. Start NX and open the model.
b. Create the gate, runner, and sprue.
c. Start analysis and show results.
d. Save the file.

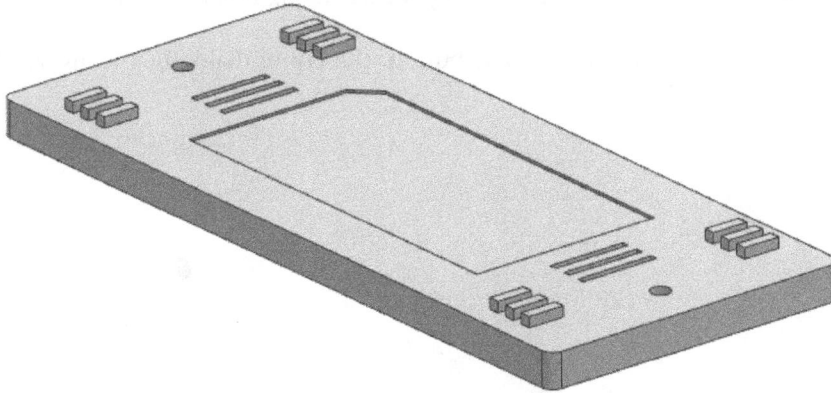

Figure 2-8 The PCB_COVER model for Tutorial 1

Starting NX and Opening the Model

First, you need to start NX and open a new file.

1. Double-click on the NX shortcut icon on the desktop of your computer to start NX.

2. Choose the **Open** button from the **Standard** group of the **Home** tab or choose **Menu > File > Open** from the **Top Border Bar**; the **Open** dialog box is displayed.

3. Select the **PCB_COVER** from the **Name** list; the **PCB_COVER** is displayed in the **File name** drop-down list. Next, choose the **OK** button; the model is displayed, refer to Figure 2-9.

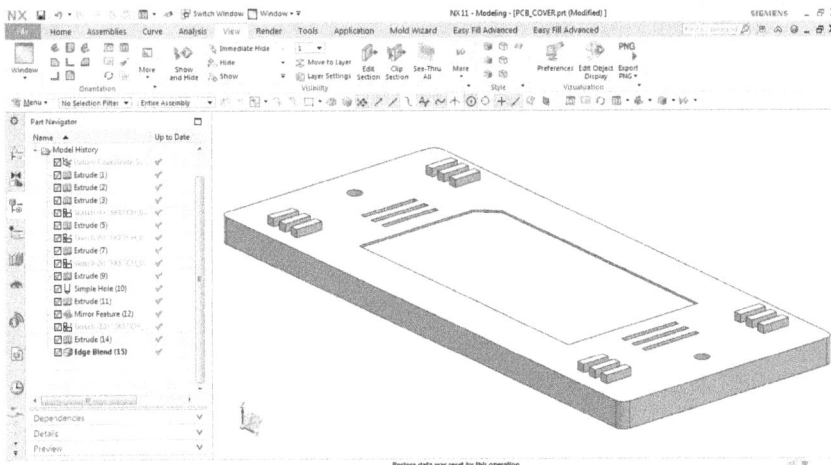

Figure 2-9 PCB_COVER model displayed in modeling environment

Set the Working Folder

In this section, you will analyze the model by using the tools in the **Easy Fill Advanced** tab. Sometimes, the material is not filled properly in the model. In such a case, you need to perform iterations by changing the gate, runner, coolant size, shape, or orientation.

1. Choose the **Easy Fill Advanced** tab. Next, choose the **Set Working Folder** tool from the **Setting** group of the tab; the **Working Folder** dialog box is displayed.

2. In this dialog box, choose the **Browse** button; the **Open** dialog box is displayed, refer to Figure 2-10.

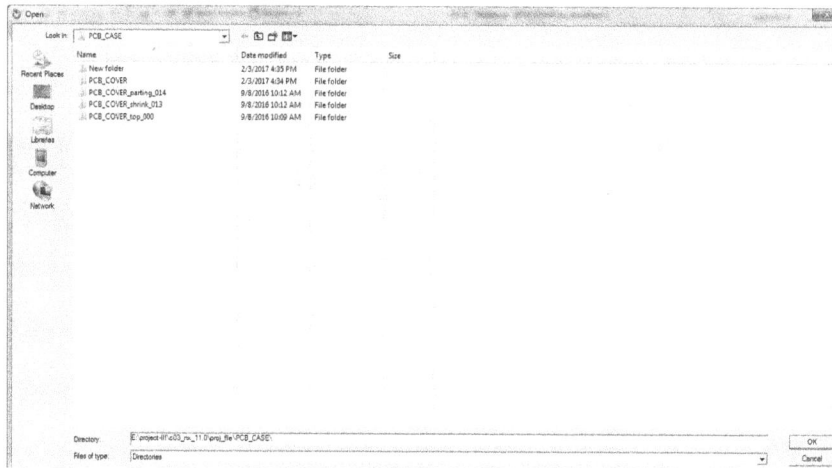

*Figure 2-10 The **Open** dialog box*

3. In this dialog box, set the working folder and choose the **OK** button from the **Open** dialog box and then from the **Working Folder** dialog box.

Creating the Gate

In this section, you will create the gate.

1. Choose the **Set Cavity** tool from the **Setting** group of the **Easy Fill Advanced** tab in the **Ribbon**; the **Select Cavity** dialog box is displayed, refer to Figure 2-11.

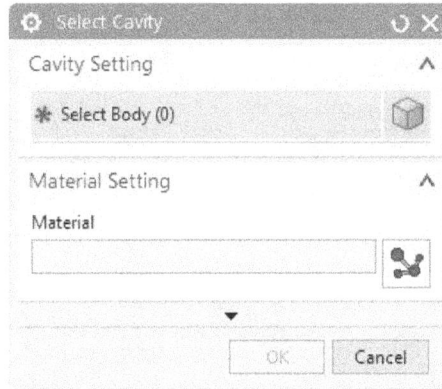

Figure 2-11 The Select Cavity dialog box

2. Select the solid body from the **Select Body** area of the **Cavity Setting** rollout to specify the solid body as cavity.

3. Choose the **Push button to select material** button in the **Material Setting** rollout; the **Moldex3D Material Wizard** dialog box is displayed, refer to Figure 2-12.

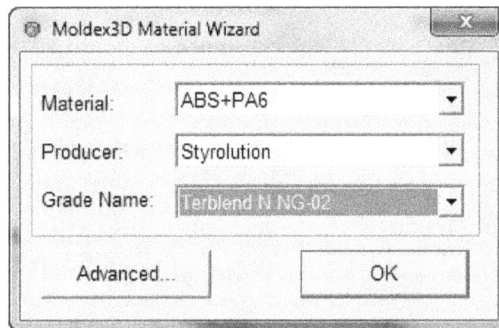

Figure 2-12 The Moldex3D Material Wizard dialog box

4. Select **ABS+PA6** from the **Material** drop-down list if not selected.

5. Select **Styrolution** from the **Producer** drop-down list if not selected.

6. Select **Terblend N NG-06** from the **Grade Name** drop-down list if not selected.

7. Choose the **OK** button from the **Moldex3D Material Wizard** and then choose the **OK** button from the **Select Cavity** dialog box.

8. Choose the **Gate Wizard** tool from the **Wizard** group; the **Create Gate** dialog box is displayed, refer to Figure 2-13.

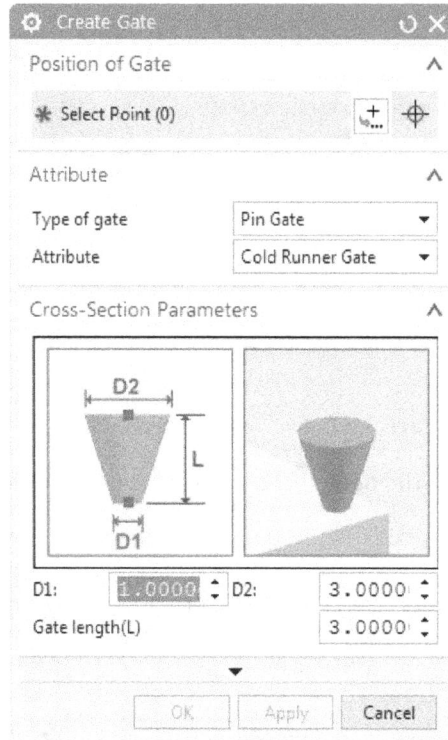

*Figure 2-13 The **Create Gate** dialog box*

9. Select the **Fan Gate** from the **Type of gate** drop-down list. Enter **3**, **3**, **8**, **4**, **180**, and **7** in **a1**, **a2**, **b1**, **b2**, **Angle**, and **L** edit boxes respectively in the **Cross-Section Parameters** rollout of the dialog box.

10. Select the **Cold Runner Gate** option from the **Attribute** drop-down list if it is not selected by default.

11. Choose the **Point Dialog** button from the **Position of Gate** rollout; the **Point** dialog box is displayed, refer to Figure 2-14.

12. Select the **Point on Face** option from the **Type** drop-down list; you are prompted to specify the point location on the desired face. The rotated view of the model indicating point location for gate placement is shown in Figure 2-15.

13. Enter **1** and **0.5** in the **U Parameter** and **V Parameter** edit boxes, respectively and choose the **OK** button to close the **Point** dialog box.

    If you need to change the dimensions of the gate then change the values in a1, a2, b1, b2, angle and gate length(L) edit boxes from the **Cross-Section Parameters** rollout, refer to Figure 2-16.

14. Choose the **OK** button to close the **Create Gate** dialog box.

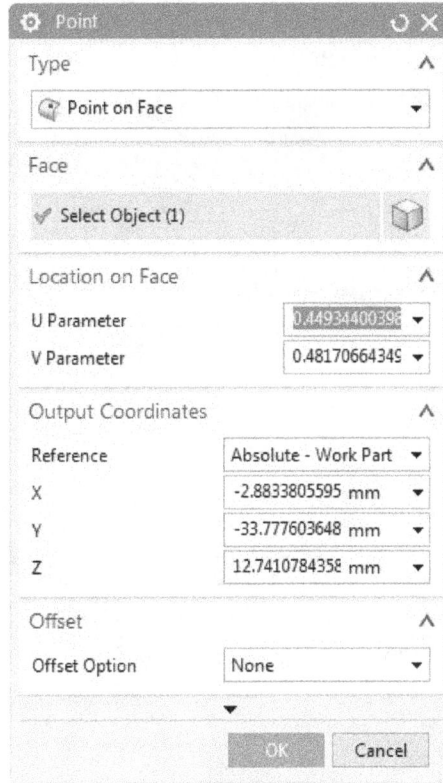

Figure 2-14 The **Point** dialog box

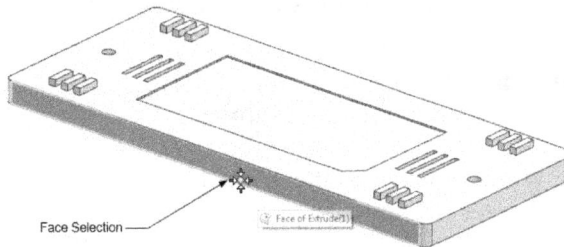

Figure 2-15 The rotated view of the model for gate placement

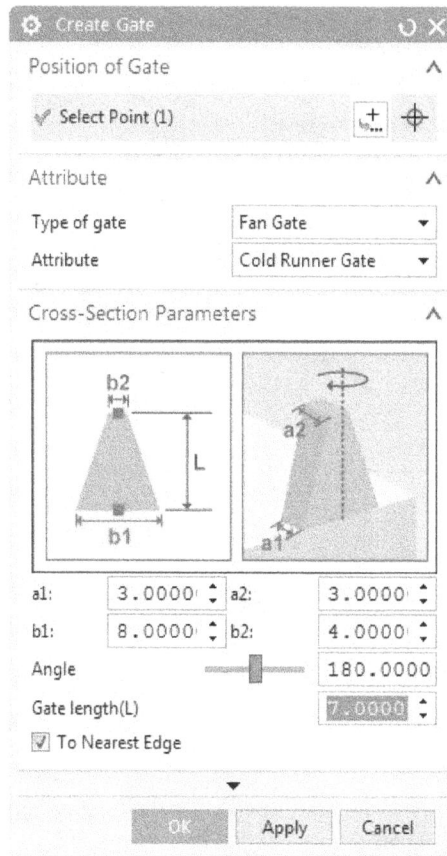

Figure 2-16 *The* *Create Gate* *dialog box*

Creating the Runner and Sprue

In this section, you will create the runner and sprue.

1. Choose the **Runner Wizard** tool from the **Wizard** group; the **Runner Wizard** dialog box is displayed, refer to Figure 2-17.

2. If the **Mold Setting** tab is not chosen by default then choose the **Reset** button.

3. Select **Z-Axis/+Z**, **2-Plate mold**, and **Cold Runner** from the **Parting Direction**, **Mold Plate Type**, and **Runner Attribute** drop-down lists, respectively if they are not selected by default.

4. Choose the **Sprue Setting** tab and then enter **10**, **6**, and **55** in the **D1**, **D2**, and **SH** edit boxes, respectively.

5. Choose the **Runner Setting** tab and select the **Circular** type of cross section for runner from the **Type** drop-down list. Enter **10** in the **D** edit box.

6. Choose the **OK** button to close this dialog box; the gate, runner, and sprue are now attached to the model, refer to Figure 2-18.

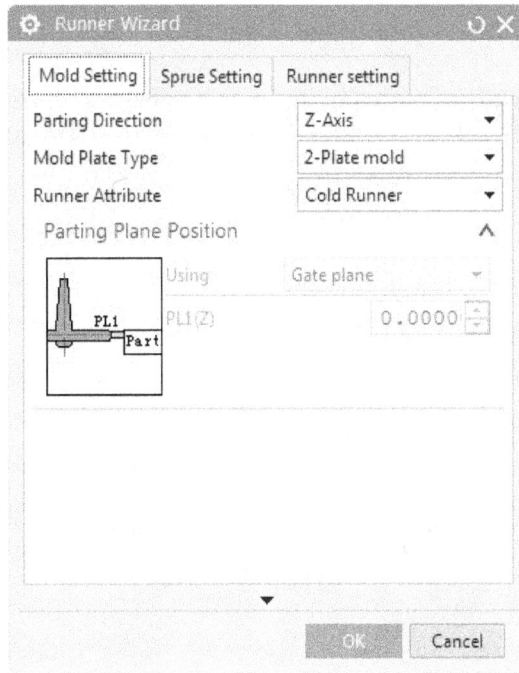

Figure 2-17 The **Runner Wizard** dialog box

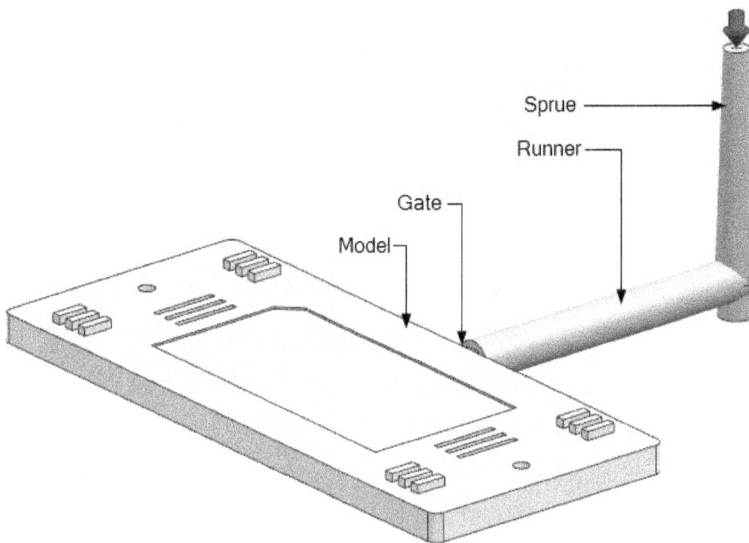

Figure 2-18 The gate, runner, and sprue attached to the model

Setting the Parting Direction

In this section, you will set the direction of ejection.

1. Choose the **Set Parting Direction** tool from the **Settings** group to set the parting direction of the model; the **Vector** dialog box is displayed, as shown in Figure 2-19.

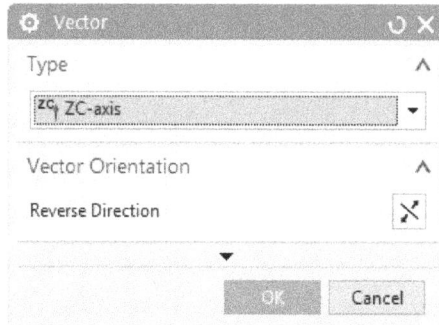

Figure 2-19   *The* ***Vector*** *dialog box*

2. Select the **ZC-axis** option from the **Type** rollout and choose the **OK** button to close the dialog box.

Starting Analysis and Analyzing Results

In this section, you will start the analysis and show results.

1. Choose the **Start Analysis** tool from the **Setting** group to start analysis of the model; the **Start Analysis** dialog box is displayed, as shown in Figure 2-20.

2. Select the **Filling & Packing - F P** option from the **Analysis** drop-down list in the **Analysis Sequence Setting** rollout.

3. Enter **220**, **0.65**, **4.5**, **260**, and **60** in the **Maximum injection pressure(MPa)**, **Filling time(Sec.)**, **Packing time(Sec.)**, **Melt Temperature(°C)**, and **Mold Temperature(°C)** edit boxes, respectively, in the **Process Condition** rollout.

4. Choose the **OK** button; the dialog box gets closed and the **Moldex3D** window showing the progress of solid mesh generation is displayed, refer to Figure 2-21.

   Note that after the solid mesh is created; the **Easy Fill Advanced Project Monitor** window showing the progress of filling and packing is displayed, refer to Figure 2-22. After completion of analysis, the **Moldex3D eDesignSYNC** window gets displayed. Choose the **OK** button.

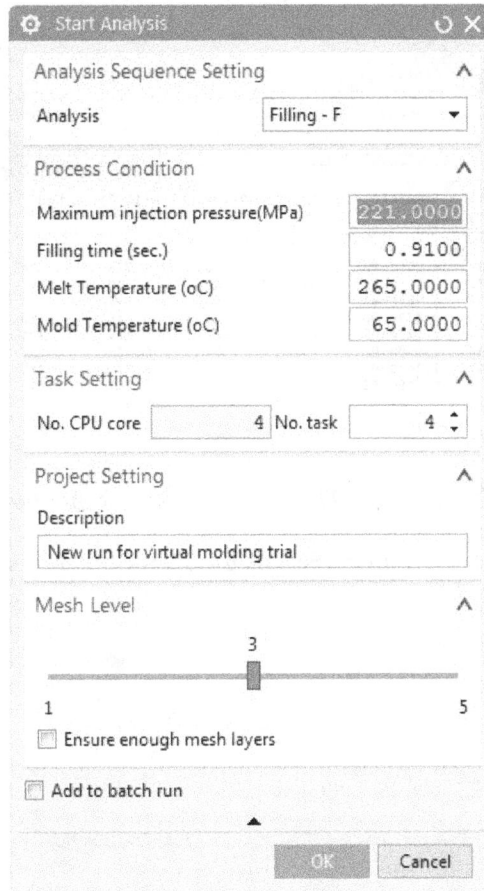

Figure 2-20 *The* **Start Analysis** *dialog box*

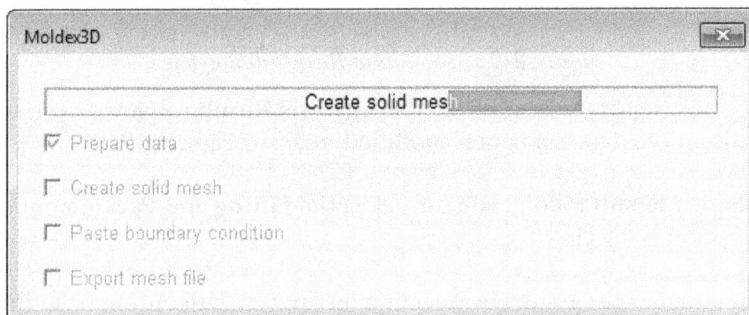

Figure 2-21 *The* **Moldex3D** *window*

5. Choose the **Close** button from the **Easy Fill Advanced Project Monitor** window.

Figure 2-22 *The **Easy Fill Advanced Project Monitor** window*

6. Choose the **Show Result** tool from the **Setting** group; the **Show Result** dialog box is displayed, refer to Figure 2-23.

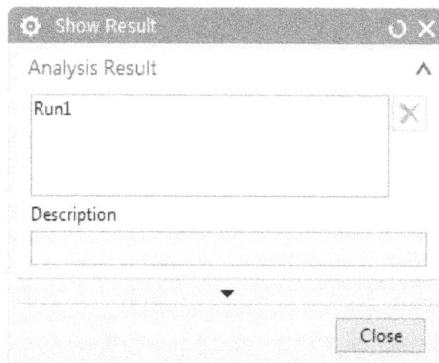

Figure 2-23 *The **Show Result** dialog box*

7. In this dialog box, select the result from the **Analysis Result** rollout. As you select the result from this rollout; the dialog box gets modified, refer to Figure 2-24.

8. From the **Display Result** rollout, you can select the **Filling and Packing** option. By default, the **Filling** option is selected.

    You can also visualize the results by selecting an appropriate option from the **Result Type** drop-down list.

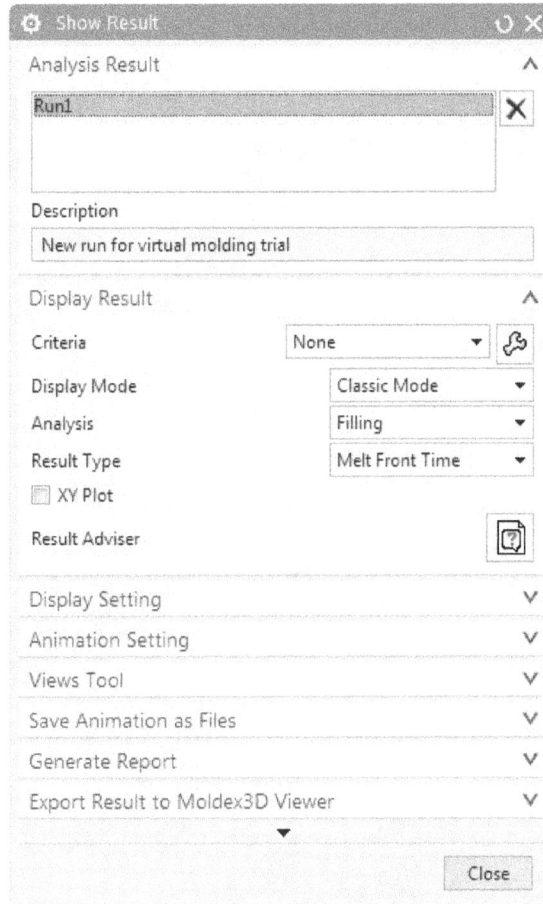

Figure 2-24 *The modified* **Show Result** *dialog box*

The options to visualize the results are **Melt Front Time**, **Air Trap**, **Weld Line**, **Volumetric Shrinkage**, **Maximum Shear Stress**, and **Sink Marks Indicator**.

9. Select the **XY Plot** check box to activate the **XY Curve Type** drop-down list. In this drop-down list, select the **Clamping Force** option; the resultant graphs are shown in Figures 2-25 through 2-30.

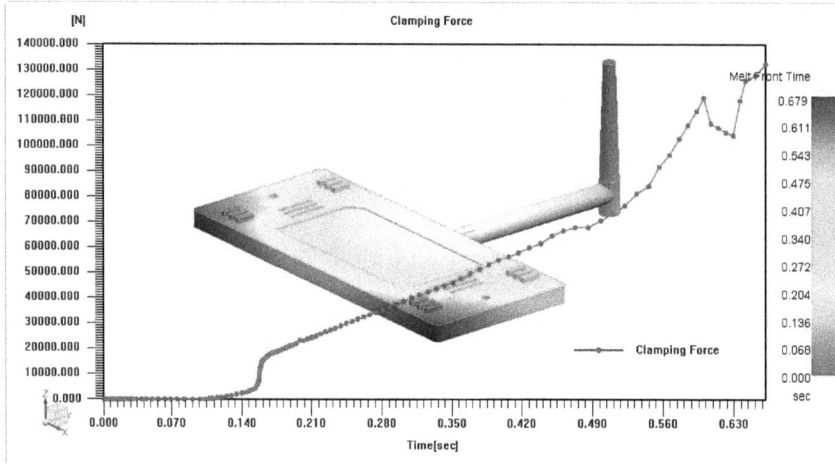

Figure 2-25 *The Clamping Force vs Melt Front Time graph*

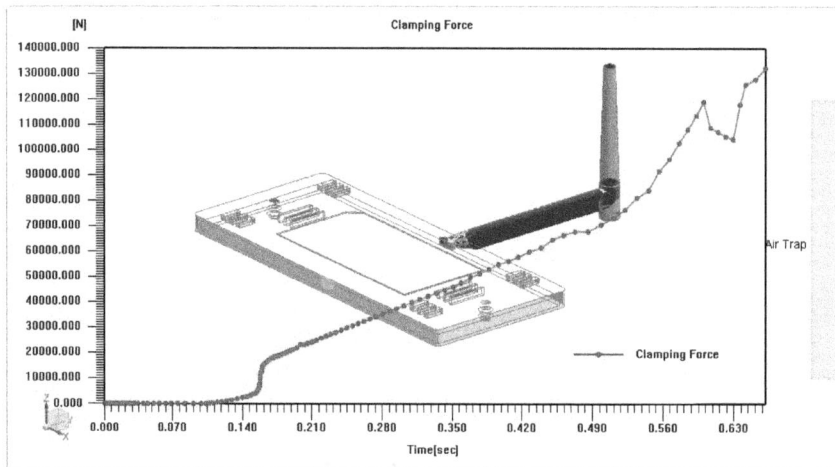

Figure 2-26 *The Clamping Force vs Air Trap graph*

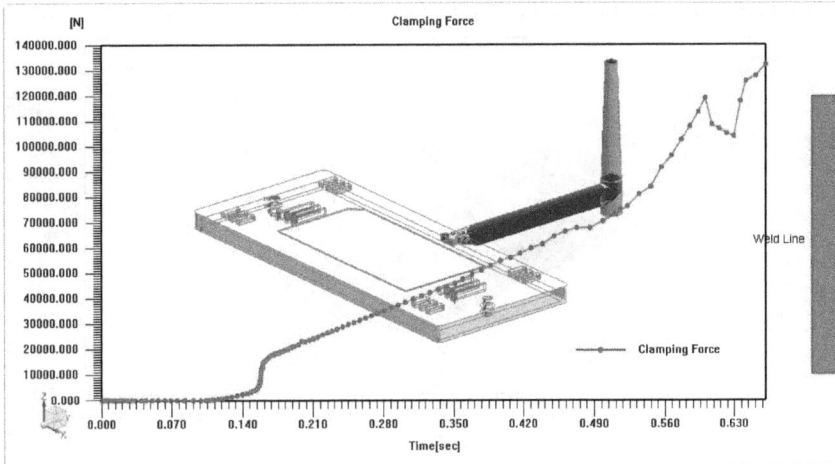

Figure 2-27 *The Clamping Force vs Weld Line graph*

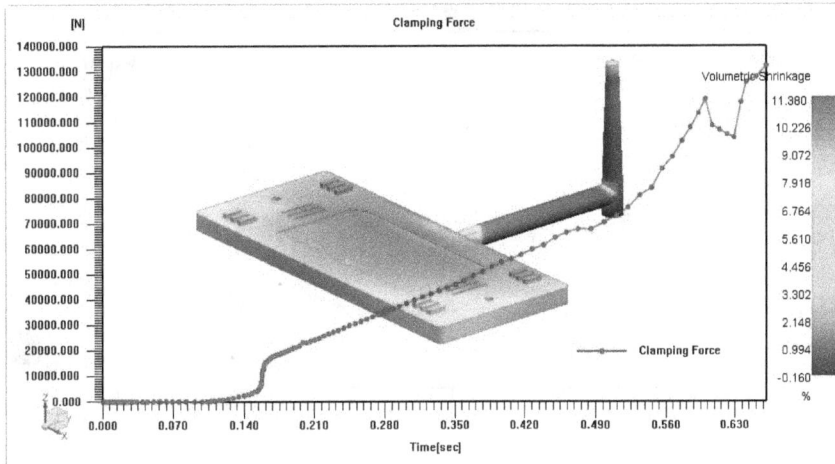

Figure 2-28 *The Clamping Force vs Volumetric Shrinkage graph*

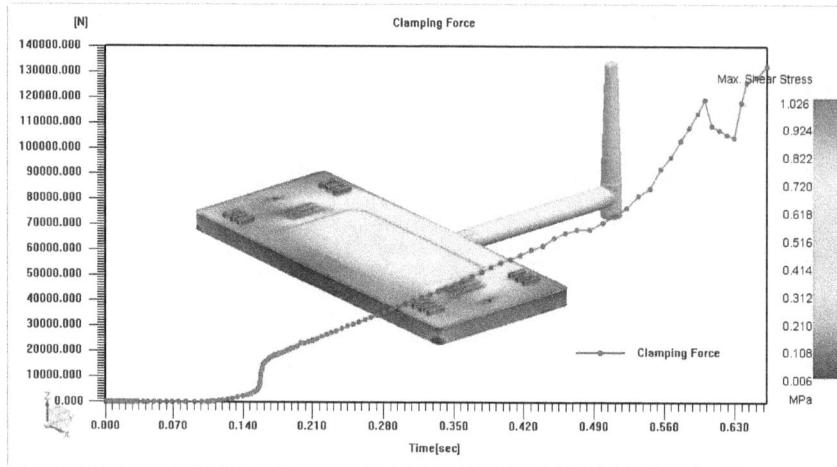

Figure 2-29 *The Clamping Force vs Max. Shear Stress graph*

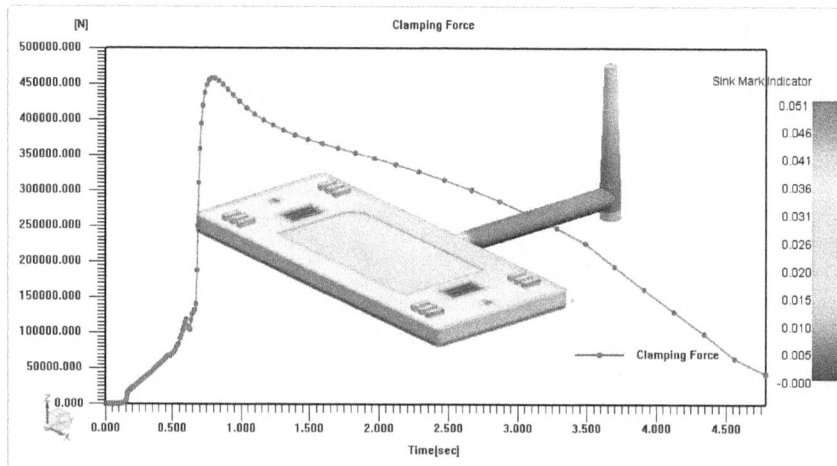

Figure 2-30 *The Clamping Force vs Sink Mark Indicator graph*

10. Choose the **Play** button from the **Animation Setting** rollout to see the flow of material. Next, choose the **Close** button to close the dialog box.

Saving and Closing the File
1. Choose **Menu > File > Close > Save and Close** from the **Top border Bar** to save and close the file.

Tutorial 2

In this tutorial, you will analyze the PCB_UPPER_CASE model which is shown in Figure 2-31. You will apply gate, runner, and sprue to analyze the flow of material into the model. You will also display and analyze the results.

After analysis, save the file with the name *c02tut2.prt* at the location given below:
            \c02   (**Expected time: 3 hr**)

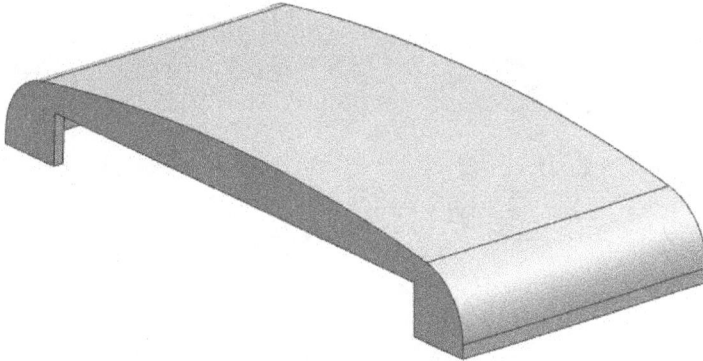

Figure 2-31 *The PCB_UPPER_CASE model for Tutorial 2*

The following steps are required to complete this tutorial:

a.   Start NX and open the model.
b.   Set a working folder.
c.   Create the gate, runner, and sprue.
d.   Start a Analysis and show results.
e.   Save the file.

Starting NX and Opening the Model

First, you need to start NX and open the file.

1.   Double-click on the shortcut icon of NX on the desktop of your computer to start NX.

2.   Choose the **Open** button from the **Standard** group of the **Home** tab or choose **Menu >
     File > Open** from the **Top Border Bar**; the **Open** dialog box is displayed.

3.   Select the **PCB_UPPER_CASE** from the **Name** list; the **PCB_UPPER_CASE** is displayed
     in **File name** drop-down list. Next, choose the **OK** button; the model is displayed, refer to
     Figure 2-32.

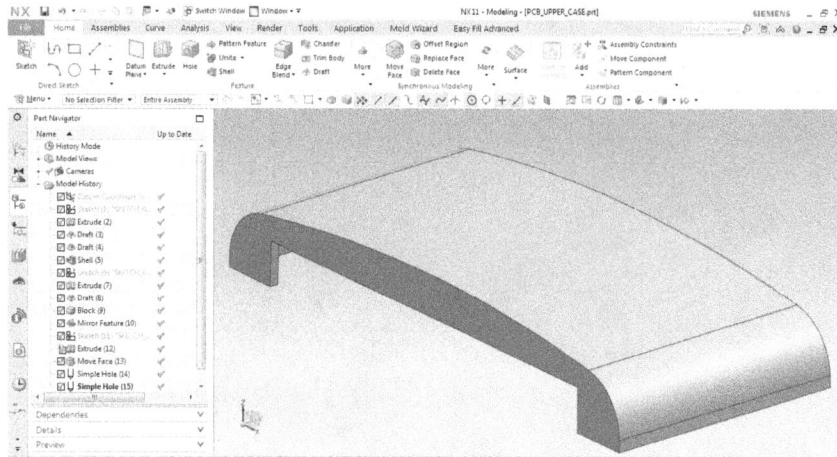

Figure 2-32 *The PCB_UPPER_CASE model in Modeling environment*

Setting the Working Folder

In this section, you will analyze the model by using the tools in the **Easy Fill Advanced** tab. Sometimes, the material is not filled properly in the model. In this case, you need to perform iterations by changing the gate, runner, coolant size, shape, or orientation.

1. Choose the **Easy Fill Advanced** tab. Next, choose the **Set Working Folder** tool from the **Setting** group of the tab; the **Working Folder** dialog box is displayed.

2. In this dialog box, choose the **Browse** button; the **Open** dialog box is displayed, refer to Figure 2-33.

Figure 2-33 *The Open dialog box*

3. Set the working folder and choose the **OK** button from the **Open** and **Working Folder** dialog boxes to close the dialog boxes.

Creating the Gate

In this section, you will create a gate.

1. Choose the **Set Cavity** tool from the **Setting** group of the **Easy Fill Advanced** tab in the **Ribbon**; the **Select Cavity** dialog box is displayed, refer to Figure 2-34.

Set Cavity

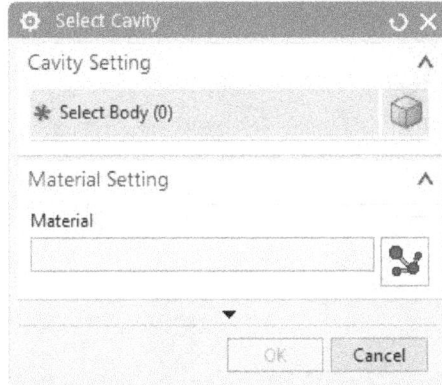

*Figure 2-34 The **Select Cavity** dialog box*

2. Select the solid body from the **Select Body** area of the **Cavity Setting** rollout to specify the solid body as cavity.

3. Choose the **Push button to select material** button in the **Material Setting** rollout; the **Moldex3D Material Wizard** dialog box is displayed, refer to Figure 2-35.

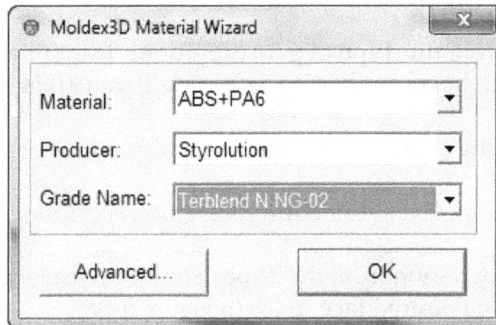

*Figure 2-35 The **Moldex3D Material Wizard** dialog box*

4. Select **ABS+PA6** from the **Material** drop-down list if it is not selected.

5. Select **Styrolution** from the **Producer** drop-down list if it is not selected.

6. Select **Terblend N NG-06** from the **Grade Name** drop-down list if it is not selected.

7. Choose the **OK** button from the **Moldex3D Material Wizard** dialog box and choose the **OK** button from the **Select Cavity** dialog box.

8. Choose the **Gate Wizard** tool from the **Wizard** group; the **Create Gate** dialog box is displayed, refer to Figure 2-36.

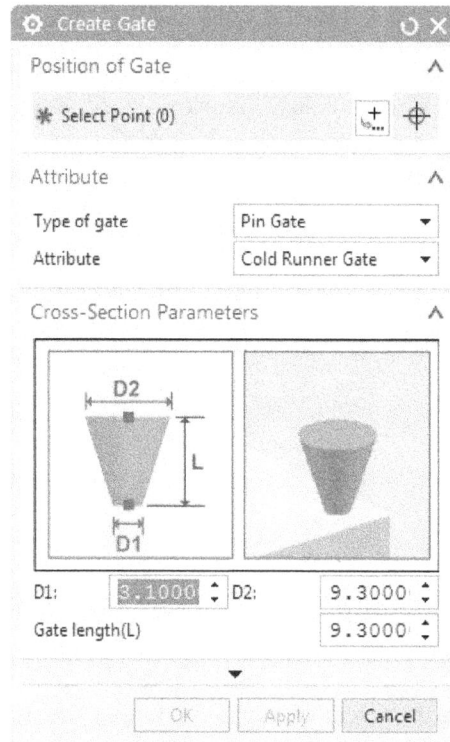

*Figure 2-36 The **Create Gate** dialog box*

9. Select the **Edge Gate** from the **Type of gate** drop-down list. Enter **3, 3, 5, 180** and **6** in **a1, a2, b, Angle**, and **L** edit boxes in the **Cross-Section Parameters** rollout of the dialog box.

10. Select **Cold Runner Gate** from the **Attribute** drop-down list if it is not selected by default.

11. Choose the **Point Dialog** button; the **Point** dialog box is displayed, refer to Figure 2-37.

12. Select the **Point on Face** option from the **Type** drop-down list; you are prompted to specify the point location on the desired face, refer to Figure 2-38.

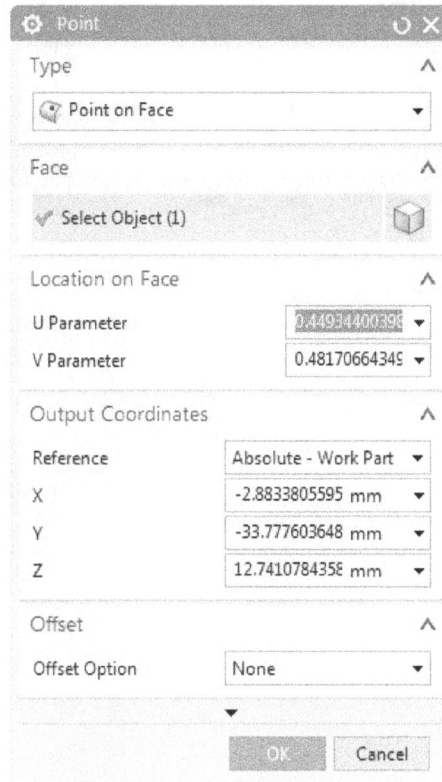

*Figure 2-37 The **Point** dialog box*

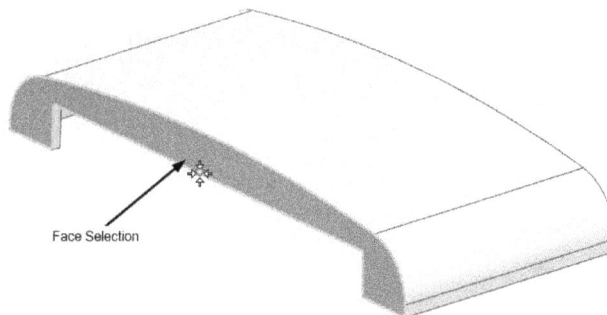

Figure 2-38 The face selected for gate placement

13. Enter **0.481** and **0.5** in the **U Parameter** and **V Parameter** edit boxes, respectively, and choose the **OK** button to close the **Point** dialog box.

    If you need to change the dimensions of the gate then select **a1**, **a2**, **b1**, **b2**, **angle** and **gate length**(L) edit box from the **Cross-Section Parameters** rollout.

14. Choose the **OK** button to close the **Create Gate** dialog box.

Creating Runner and Sprue for Analyzing the Model

In this section, you will create the runner and sprue.

1. Choose the **Runner Wizard** tool from the **Wizard** group; the **Runner Wizard** dialog box is displayed, refer to Figure 2-39.

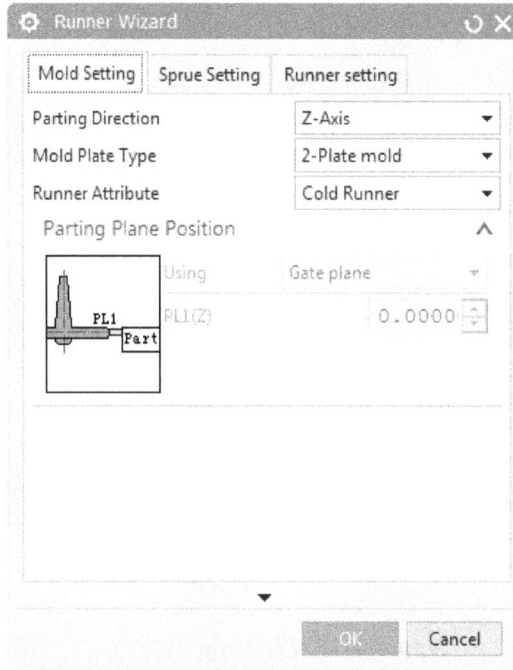

*Figure 2-39 The **Runner Wizard** dialog box*

2. By default, the **Mold Setting** tab is chosen. If it is not chosen by default then the use **Reset** button in dialog box.

3. Select **Z-Axis/+Z**, **2-Plate mold**, and **Cold Runner** from the **Parting Direction**, **Mold Plate Type**, and **Runner Attribute** drop-down lists, respectively if not selected by default.

4. Choose the **Sprue Setting** tab and then enter **10**, **6**, and **55** in the **D1**, **D2**, and **SH** edit boxes, respectively.

5. Choose the **Runner Setting** tab and select the **Circular** type of cross section for runner from the **Type** drop-down list. Enter **10** in the **D** edit box.

6. Choose the **OK** button to close this dialog box; the gate, runner and sprue are now attached to the model, refer to Figure 2-40.

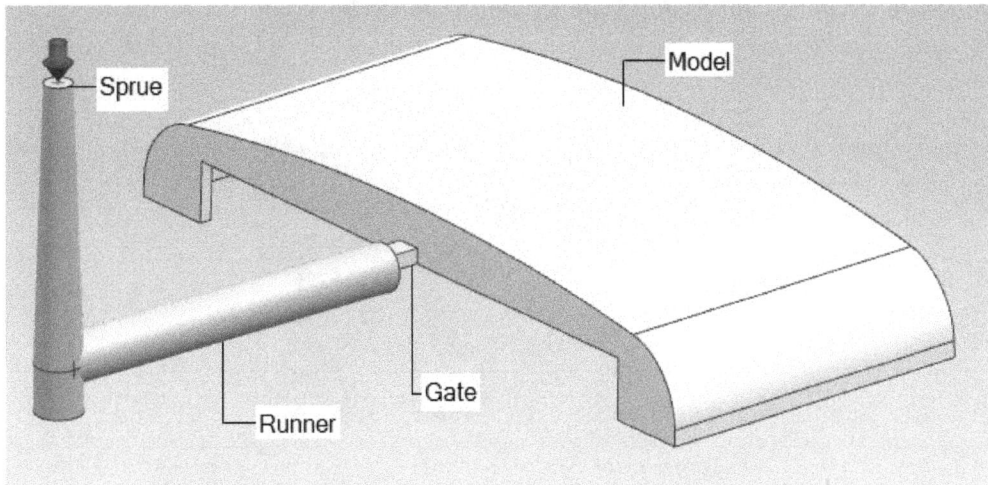

Figure 2-40 *The gate, runner, and sprue attached to the model*

Setting the Parting Direction

To analyze the model, you need to set up the parting direction.

1. Choose the **Set Parting Direction** tool from the **Settings** group to set the parting direction of the model; the **Vector** dialog box is displayed, refer to Figure 2-41.

2. Select the **ZC-axis** option from the **Type** rollout and choose the **OK** button to close the dialog box.

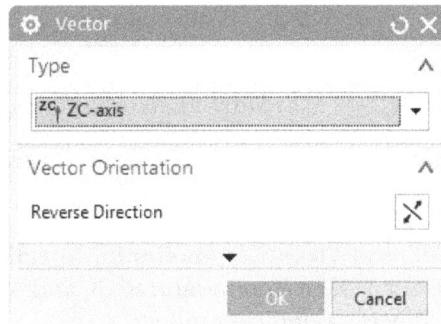

Figure 2-41 *The **Vector** dialog box*

Starting Analysis and Analyzing Results

In this section, you will start the analysis and show results.

1. Choose the **Start Analysis** tool from the **Setting** group to start the analysis of the model; the **Start Analysis** dialog box is displayed, as shown in Figure 2-42.

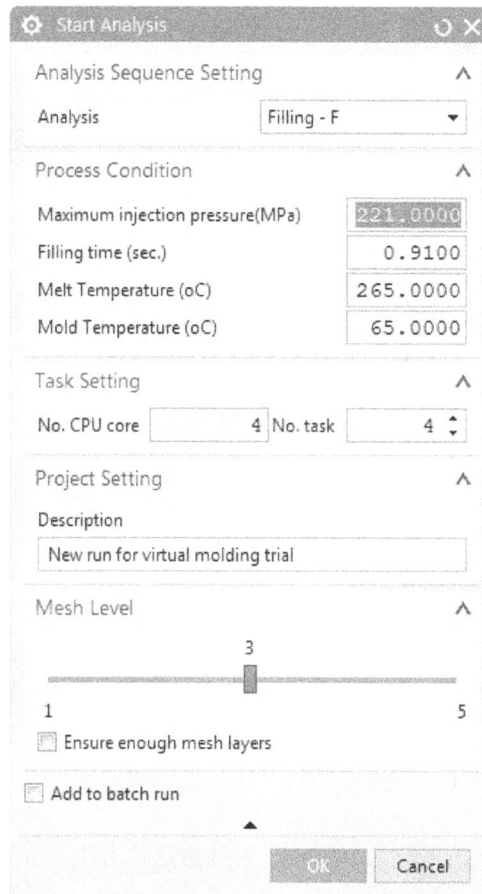

*Figure 2-42 The **Start Analysis** dialog box*

2. Select the **Filling & Packing - F P** option from the **Analysis** drop-down list of the **Analysis Sequence Setting** rollout.

3. Enter **165**, **0.58**, **6.5**, **270**, and **70** in the **Maximum injection pressure(MPa)**, **Filling time(Sec.)**, **Packing time(Sec.)**, **Melt Temperature(°C)**, and **Mold Temperature (°C)** edit boxes respectively in the **Process Condition** rollout.

4. Choose the **OK** button; the dialog box gets closed and the **Moldex3D** window showing the progress of solid mesh generation is displayed, refer to Figure 2-43.

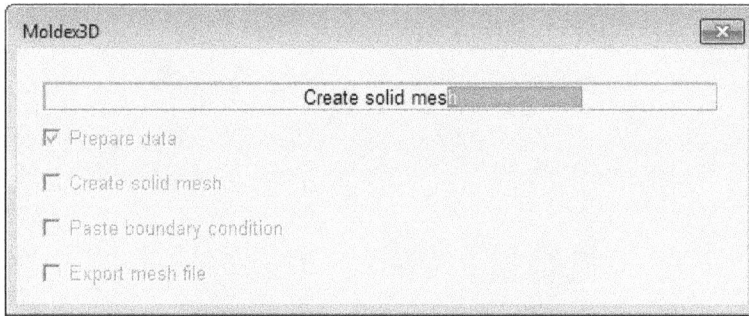

*Figure 2-43 The **Moldex3D** window*

Note that after the solid mesh is created; the **Easy Fill Advanced Project Monitor** window showing the progress of filling and packing is displayed, refer to Figure 2-44. After completion of analysis, the **Moldex3D eDesignSYNC** window gets displayed. Choose the **OK** button.

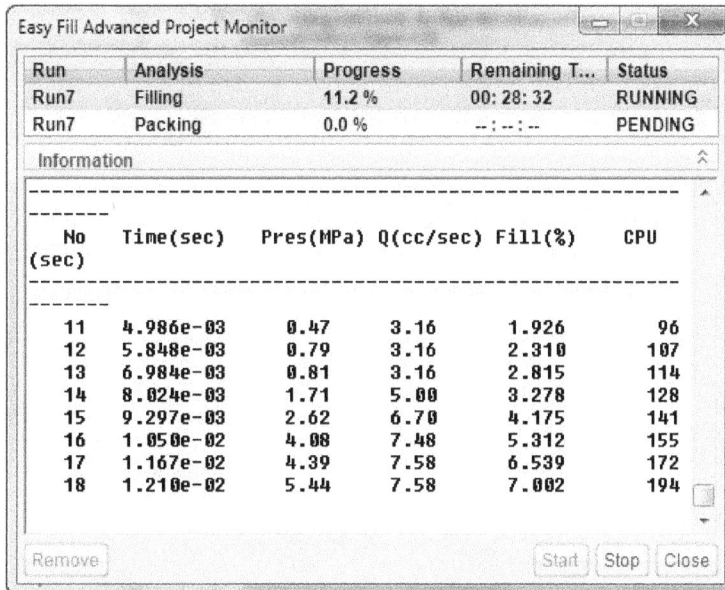

*Figure 2-44 The **Easy Fill Advanced Project Monitor** window*

5. Choose the **Close** button from the **Easy Fill Advanced Project Monitor** window.

6. Choose the **Show Result** tool from the **Settings** group; the **Show Result** dialog box is displayed, refer to Figure 2-45.

7. Select result from the **Analysis Result** rollout; the dialog box gets modified, refer to Figure 2-46.

   From the **Display Result** rollout, you can select the **Filling and Packing** option. By default, the **Filling** option is selected.

You can select the **Filling** or **Packing** option from the **Analysis** drop-down list in the **Display Result** rollout. By default, the **Filling** option is selected.

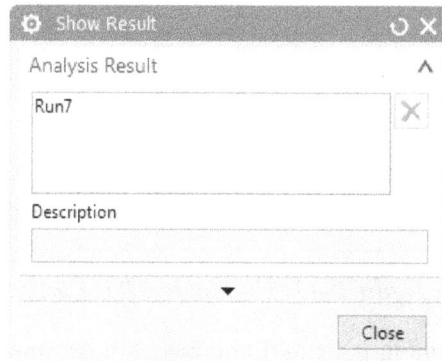

*Figure 2-45 The **Show Result** dialog box*

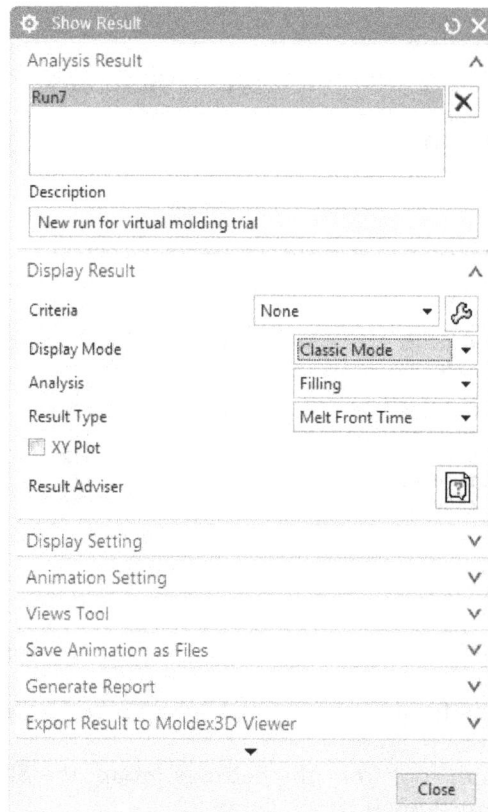

*Figure 2-46 The modified **Show Result** dialog box*

In this tutorial, you can visualize the results by selecting appropriate option from the **Result Type** drop-down list.

The options which can be selected for results are **Melt Front Time**, **Air Trap**, **Weld Line**, **Volumetric Shrinkage**, **Maximum Shear Stress**, and **Sink Marks Indicator**.

8. Select the **XY Plot** check box to activate the **XY Curve Type** drop-down list. Next, select the **Clamping Force** option from this drop-down list; the resultant graphs are shown, refer to Figure 2-47 through Figure 2-52.

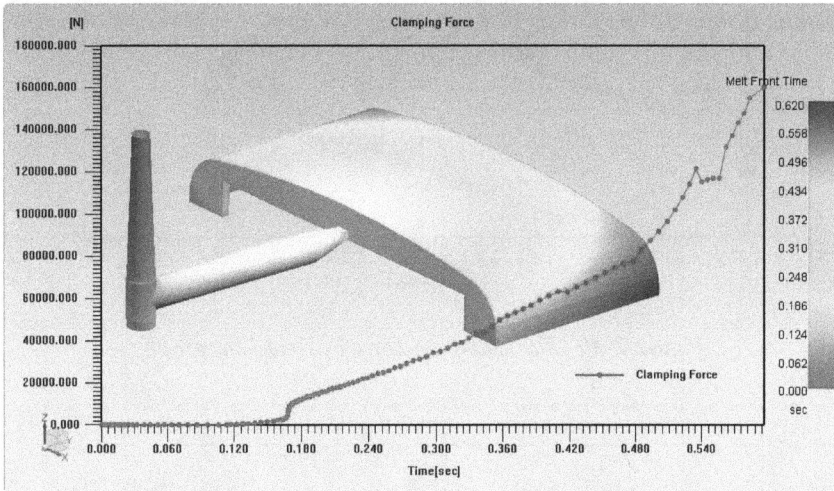

Figure 2-47 *The Clamping Force vs Melt Front Time graph*

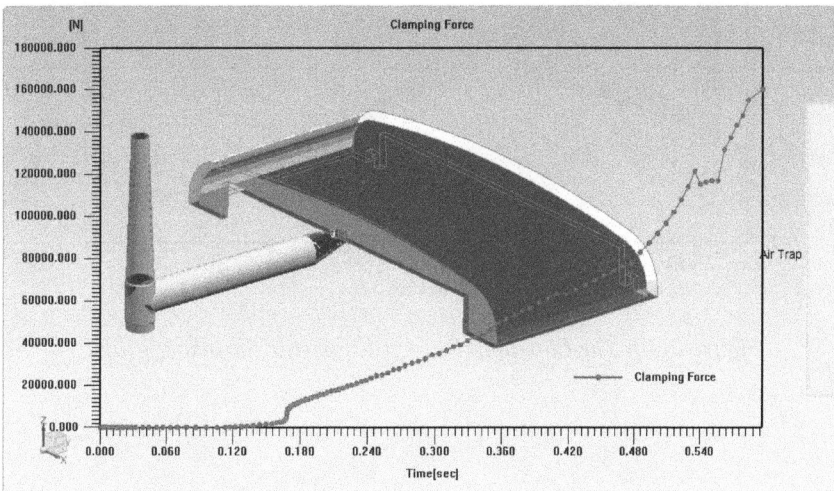

Figure 2-48 *The Clamping Force vs Air Trap graph*

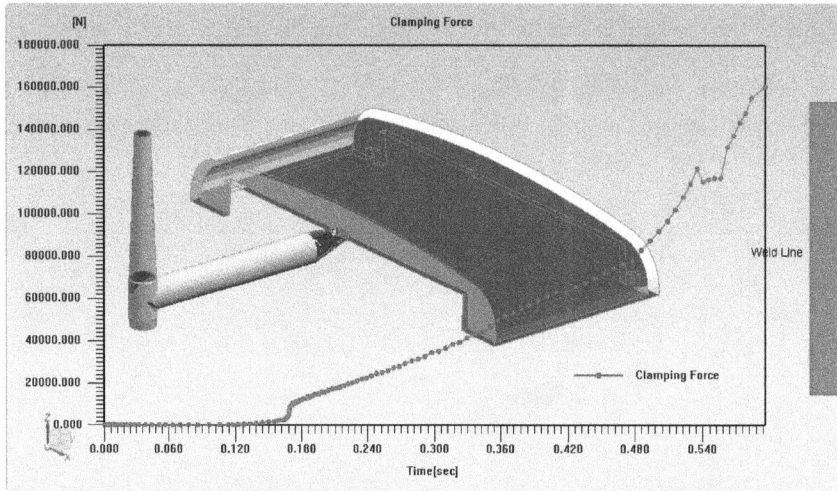

Figure 2-49 The Clamping Force vs Weld Line graph

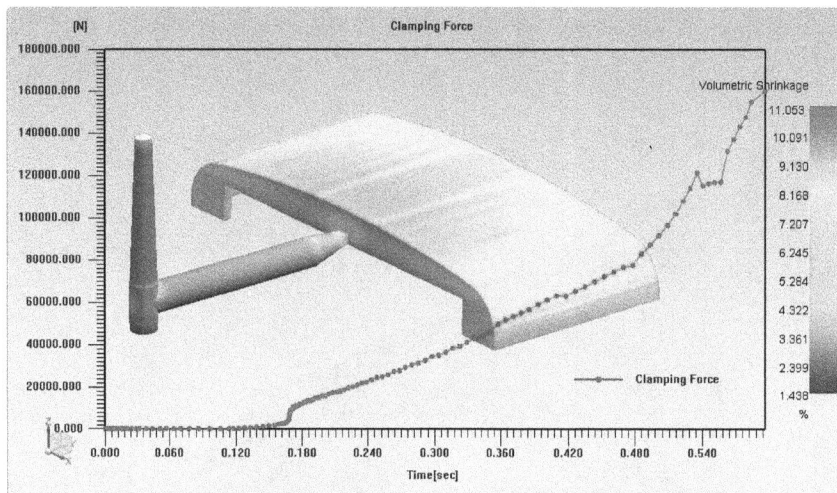

Figure 2-50 The Clamping Force vs Volumetric Shrinkage graph

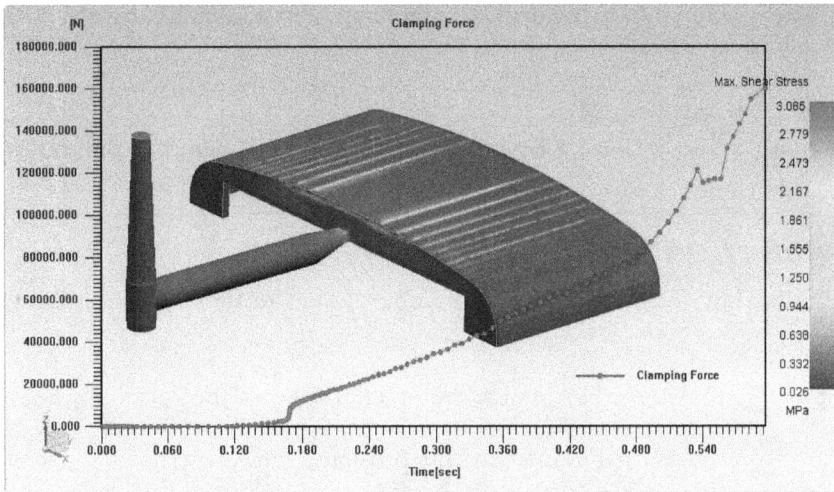

Figure 2-51 The Clamping Force vs Max. Shear Stress graph

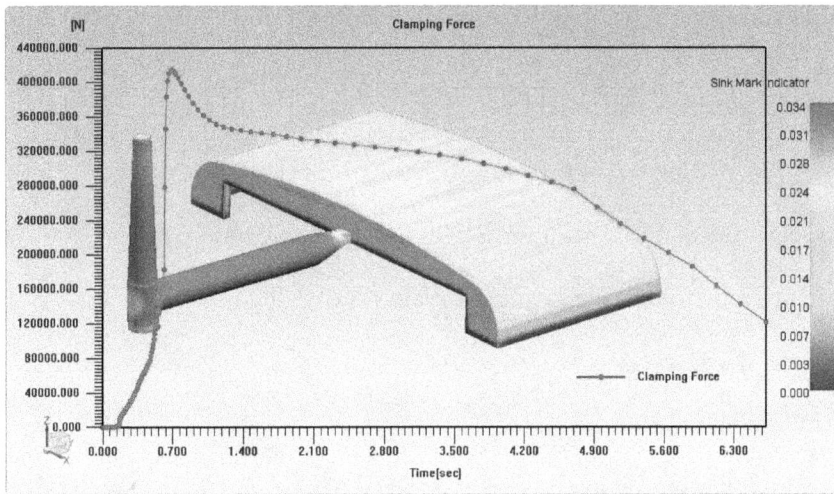

Figure 2-52 The Clamping Force vs Sink Mark Indicator graph

9. Choose the **Play** button from the **Animation Setting** rollout. Next, choose the **Close** button to close the dialog box.

Saving and Closing the File
1. Choose **Menu > File > Close > Save and Close** from the **Top border Bar** to save and close the file.

Self-Evaluation Test

Answer the following questions and then compare them to those given at the end of this chapter:

1. The _____ tool is used to analyze the wall thickness of the part.

2. The _____ tool is used to calculate the projected area of a part.

3. Clamping Force is calculated by multiplying injection pressure and _____.

4. The _____ window is used to show the projected area value.

5. You can use the _____ or _____ option to calculate the wall thickness in the **Check Wall Thickness** dialog box.

6. The _____ dialog box is used to check the product quality.

7. You can select the material from the **Select Cavity** dialog box. (T/F)

8. There are two types of attributes in the **Create Gate** dialog box. (T/F)

Review Questions

Answer the following questions:

1. Which of the following dialog boxes is displayed when you choose the **Set Cavity** button from the **Easy Fill Advanced** tab to specify the cavity of the component?

   (a) **Set Cavity**   (b) **Select Cavity**
   (c) **Cavity**   (d) **Set**

2. Which of the following tools in NX Mold Wizard is used to start analysis in the **Easy Fill Advanced** tab?

   (a) **Analysis**   (b) **Start**
   (c) **Start Analysis**   (d) None of these

3. You can select the _____ and _____ option of analysis, from the Display Result rollout of the Show Result dialog box.

4. You can use the _____ area to estimate the required ram force for EDM machining.

5. You need not to specify the plane while calculating the projected area using the **Calculate Area** tool. (T/F)

EXERCISE

To perform the exercise, you need to download the zipped file named as *c02_NX_Mold_input* from the **Input Files** section of the CADCIM website. The complete path for downloading the file is:

> *Textbooks > CAD/CAM > NX_Mold > Mold Design using NX 11.0: A Tutorial Approach > Input Files*

After the file is downloaded, extract the folder.

Exercise 1

In this exercise, you need to open the model that you have downloaded and then create and position gate, runner, and sprue and analyze the following results. Figure 2-53 shows the model for this exercise.

(a) Clamping Force vs Melt Front Time graph
(b) Clamping Force vs Air Trap graph
(c) Clamping Force vs Weld Line graph
(d) Clamping Force vs Volumetric Shrinkage graph
(e) Clamping Force vs Max. Shear Stress graph
(f) Clamping Force vs Sink Mark Indicator graph

(Expected time: 70 min)

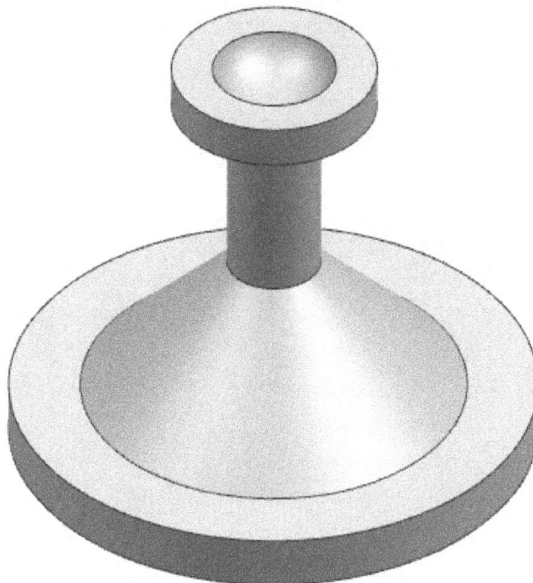

Figure 2-53 Model for Exercise 1

Answers to Self-Evaluation Test

1. Check Wall Thickness, **2.** Calculate Area, **3.** projected area, **4.** Information, **5.** Ray, Rolling ball, **6.** Mold Design Validation, **7.** T, **8.** T

Chapter 3

Creating Parting Surface

Learning Objectives

After completing this chapter, you will be able to:

- *Initialize a project*
- *Define Mold CSYS*
- *Check core and cavity region*
- *Create a patch surface*
- *Define core and cavity region*
- *Create a parting surface*

INTRODUCTION

In the previous chapters, you have learned part analysis. In this chapter, you will learn to create parting surface. Also, you will learn to use tools for creating parting surface.

PARTING SURFACE

Parting surface is a surface feature that you can use to split any mold workpiece. The relationship between a parting surface and a component in NX Mold is shown in Figure 3-1.

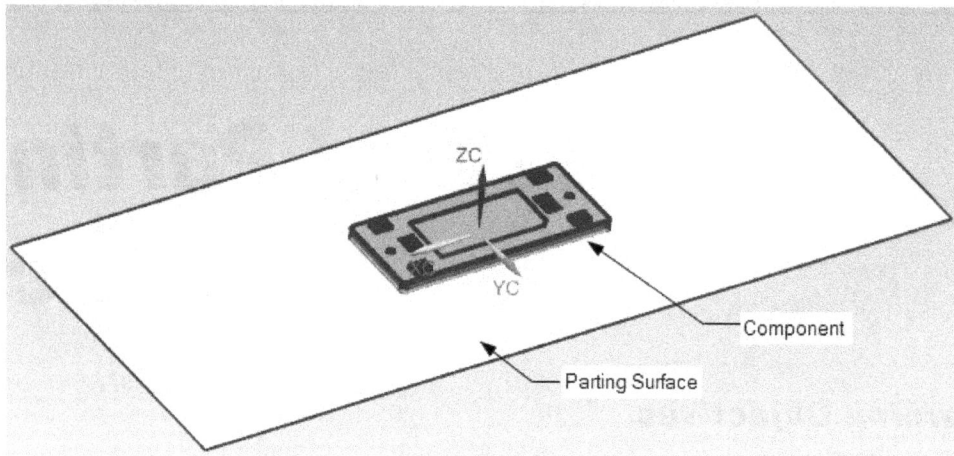

Figure 3-1 *The parting surface in NX Mold*

A Parting surface is classified into two categories - flat and non-flat parting surface. The non-flat parting surface includes stepped, profiled, and angled parting surfaces.

A flat parting surface is used when two parting surfaces are not perfectly matched and the molten material from the impression is leaking through the gap. This leaking material is called flash. To avoid flash, it is recommended to use a flat parting surface near the boundary of mold volumes. Generally, flat parting surfaces are easy to manufacture and maintain.

You need to design parting surface in such a way the component can be easily removed from the mold. NX mold wizard provides you to automatically create parting surface and helps to analyze the part so that you can place the parting surface in a correct manner.

TYPES OF PARTING SURFACE

1) Flat parting surface
2) Non-flat parting surface

- Stepped parting surface
- Profiled parting surface
- Angled parting surface
- Complex edge forms surface
- Local stepped parting surface

For creating a parting surface in NX mold wizard, you need to understand the tools used for creating parting surfaces that are available in the **Parting Tools** gallery. Some of these tools are listed below:

- Check Regions
- Patch Surface
- Define Regions
- Design Parting Surface

Before creating parting surface, you should know how to load and apply material to the component. You should also know how to place the model according to the injection molding machine.

INITIALIZE PROJECT

Ribbon:	Mold Wizard > Initialize Project

The **Initialize Project** tool is used to create a new mold design project or to add new model to create a family mold. To do so, choose the **Initialize Project** tool of the **Mold Wizard** tab; the **Initialize Project** dialog box will be displayed, as shown in Figure 3-2.

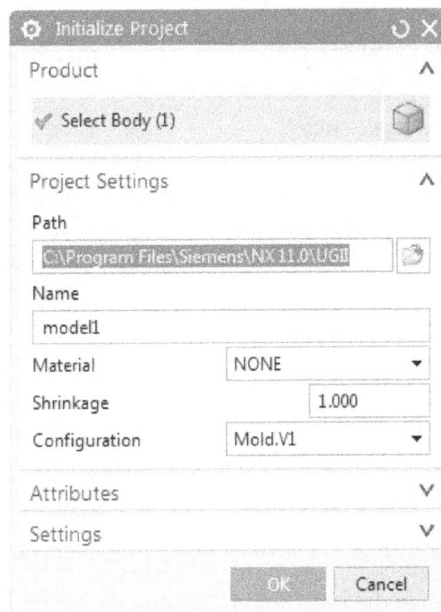

*Figure 3-2 The **Initialize Project** dialog box*

In the **Path** edit box in this dialog box, you can type the path or choose the **Browse** button to define the path for saving the file. Select the type of material from the **Material** drop-down list; the shrinkage value gets changed in the **Shrinkage** edit box according to the type of material selected.

You can select the types of template such as **Mold.V1**, **ESI**, and **Original** from the **Configuration** drop-down list.

You can also define customer name, designer name, project number, and project description in the **Attributes** rollout which will be available on expanding the dialog box.

You can select units of measure for the tool assembly files from the **Project Units** drop-down list of the **Settings** rollout.

MOLD CSYS

Ribbon: Mold Wizard > Main gallery > Mold CSYS

You can reorient the original component according to the mold assembly or the injection molding machine. You should consider following points while placing the component in mold wizard:
  • Orient the component such that ejection occurs in z-axis of the mold base.
  • Position the component in such a way that the parting plane lies on the X-Y plane.

To orient the component, invoke the **Mold CSYS** tool from the **Main** gallery of the **Mold Wizard** tab; the **MOLD CSYS** dialog box will be displayed, refer to Figure 3-3.

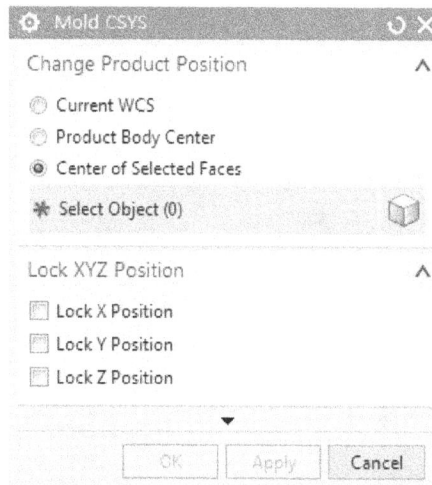

Figure 3-3 The MOLD CSYS dialog box

There are three options to change the position of the component in the **Change Product Position** rollout. These options are discussed next.

Current WCS

It helps you to reposition and reorient a component according to the position and orientation of the mold base.

Product Body Center

This radio button is used to reposition the WCS at the center of the component body. When you select this radio button, the **Lock XYZ Position** rollout will be displayed. The options in this rollout help you to lock the position of X,Y, and Z coordinates of the component.

Center of Selected Faces

This option helps to reposition the component from the center of one or more selected faces to the mold base origin. As you select the **Center of Selected Faces** radio button, the **Lock XYZ Position** rollout gets displayed. The options in this rollout help to lock the position of X,Y, and Z coordinates of the component.

CHECK REGIONS

Ribbon: Mold Wizard > Parting Tools gallery > Check Regions

The **Check Regions** tool is used to analyze the draft angle of faces, set the color in the faces according to the draft values, and find undercut in the model. It also helps to find the core and cavity regions in the model. To analyze the core and cavity region of the model, invoke this tool from the **Parting Tools** gallery of the **Mold Wizard** tab; the **Check Regions** dialog box and the **Parting Navigator** tree will be displayed, refer to Figures 3-4 (a) and (b).

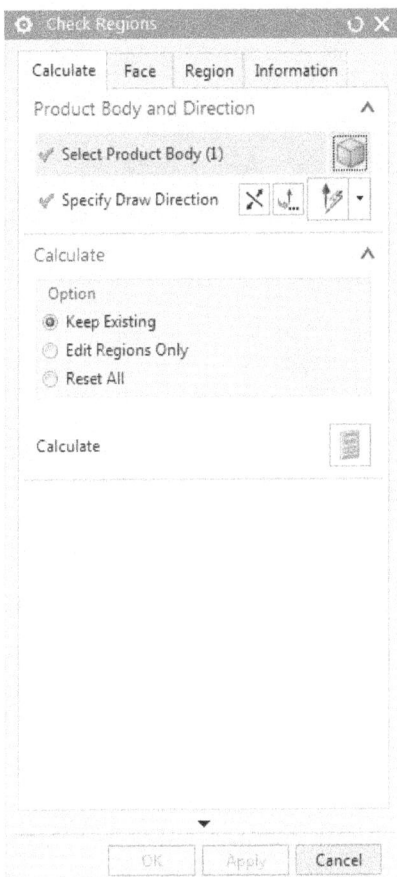

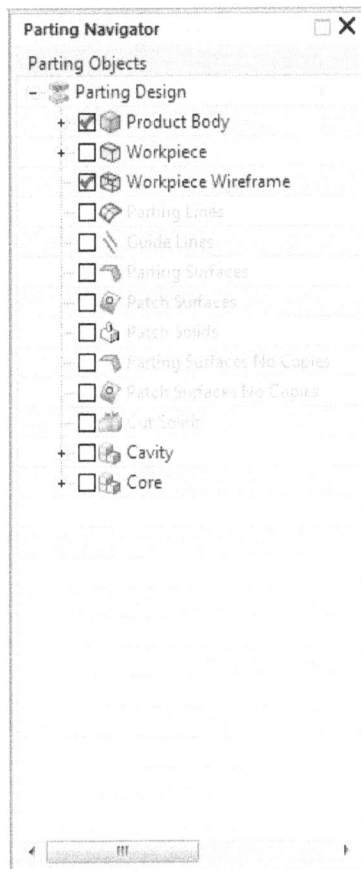

*Figure 3-4 (a) The **Check Regions** dialog box*   *Figure 3-4 (b) The **Parting Navigator** tree*

PATCH SURFACE

Ribbon: Mold Wizard > Parting Tools > Patch Surface

The **Patch Surface** tool is used to create sheets to close the openings in the model. To patch a surface, invoke the **Patch Surface** tool from the **Parting Tools** gallery of the **Mold Wizard** tab; the **Edge Patch** dialog box will be displayed, refer to Figure 3-5. In NX Mold Wizard, there are three options in the **Type** drop-down list of the **Loop Selection** rollout: **Face**, **Body**, and **Traverse**. The procedure to create patch surface will be discussed in the tutorials of this chapter.

DEFINE REGIONS

Ribbon: Mold Wizard > Parting Tools > Define
               Regions

The **Define Regions** tool is used to create core and cavity region sheets, parting lines, and other region sheets (for sliders and lifters). To create core and cavity region sheets, invoke the **Define Regions** tool from the **Parting Tools** gallery of the **Mold Wizard** tab; the **Define Region**s dialog box will be displayed, refer to Figure 3-6.

In the **Define Regions** rollout of this dialog box, you can examine the number of faces selected for cavity, core, and undefined faces. The procedure to create regions for core and cavity will be discussed in the tutorials of this chapter.

DESIGN PARTING SURFACE

Ribbon: Mold Wizard > Parting Tools > Design
               Parting Surface

The **Design Parting Surface** tool is used to create a parting surface in the mold component. To create a parting surface, invoke the **Design**

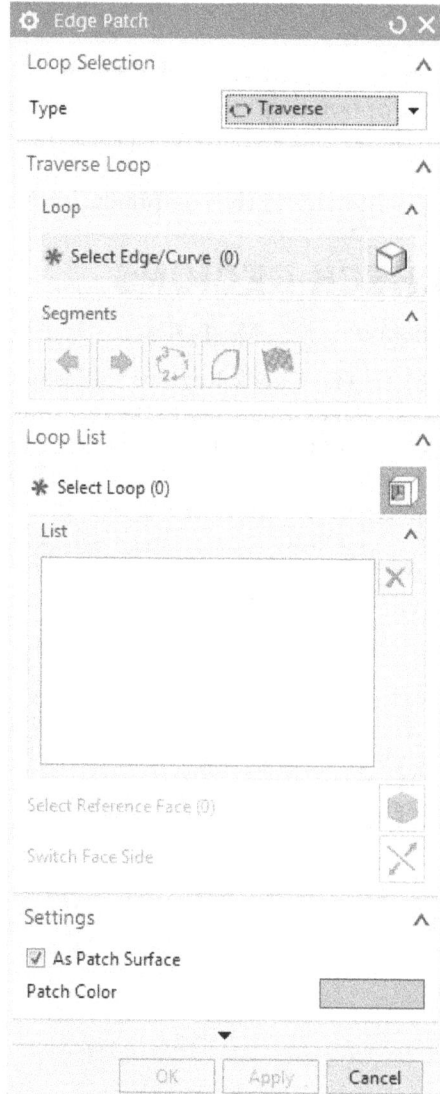

*Figure 3-5 The **Edge Patch** dialog box*

Parting Surface tool from the **Parting Tools** gallery of the **Mold Wizard** tab; the **Design Parting Surface** dialog box will be displayed, refer to Figure 3-7. You can use the tools available in the **Create Parting Surface** rollout to create the parting surfaces. The tools available in this rollout are: **Extrude**, **Swept**, **Enlarged Surface**, **Bounded Plane**, **Extend Sheet**, **Ribbon Surface**, and **Guided Extension**.

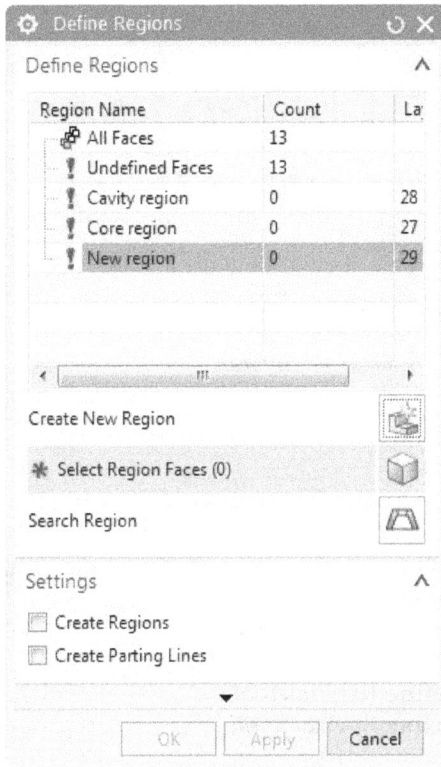

Figure 3-6 *The* **Define Regions** *dialog box*

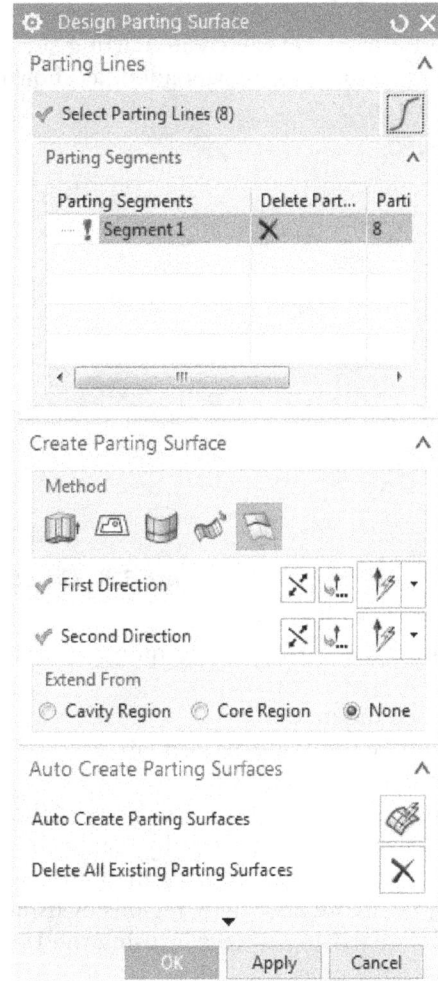

Figure 3-7 *The* **Design Parting Surface** *dialog box*

TUTORIALS

In the tutorials, you will use the parting surface tools for creating the parting surface. To perform the tutorials, you need to download the zipped file named as *c03_NX_Mold_input* from the **Input Files** section of the CADCIM website. The complete path for downloading the file is given below:

> *Textbooks > CAD/CAM > NX_Mold > Mold Design using NX 11.0: A Tutorial Approach > Input Files*

After the file is downloaded, extract the folder. In this folder, you will find Tut1 and Tut2 folders containing input files for Tutorial 1 and Tutorial 2.

Tutorial 1

In this tutorial, you will create parting surface for the PCB_COVER model contained in Tut 1 folder that you have downloaded, as shown in Figure 3-8. After creating the parting surface, save the file.   (**Expected time: 1 hr**)

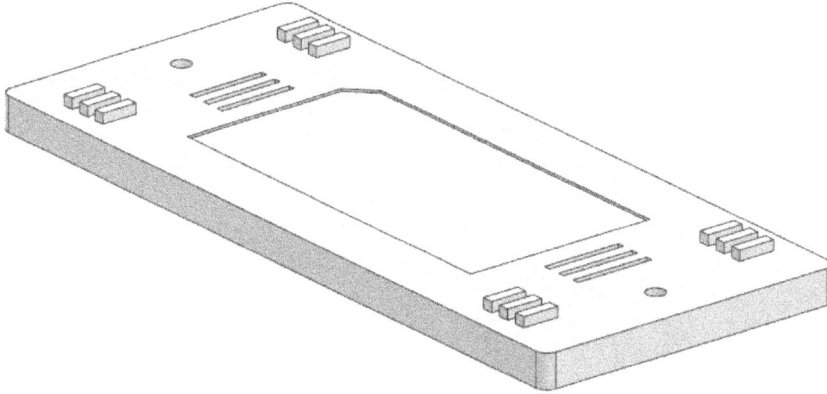

Figure 3-8 *The PCB_COVER model for Tutorial 1*

The following steps are required to complete this tutorial:

a.   Open the model.
b.   Choose the **Mold Wizard** tab.
c.   Initialize the project.
d.   Reorient the model by using the **Mold CSYS** tool.
e.   Analyze the core and cavity regions by using the **Check Regions** tool.
f.   Patch the surface by using the **Patch Surface** tool.
g.   Specify core and cavity regions by using the **Define Regions** tool.
h.   Create parting surface by using the **Design Parting Surface** tool.
i.   Save the model.

Starting NX and Opening a Model

First, you need to start NX and then open a new file.

1.   Double-click on the shortcut icon of NX available on the desktop of your computer to start NX.

2.   Choose the **Open** button from the **Standard** group of the **Home** tab or choose **Menu > File > Open** from the **Top Border Bar**; the **Open** dialog box is displayed.

3. Select the **PCB_COVER** from the **Name** list; **PCB_COVER** is displayed in the **File name** drop-down list. Then, choose the **OK** button; the model is displayed, refer to Figure 3-9.

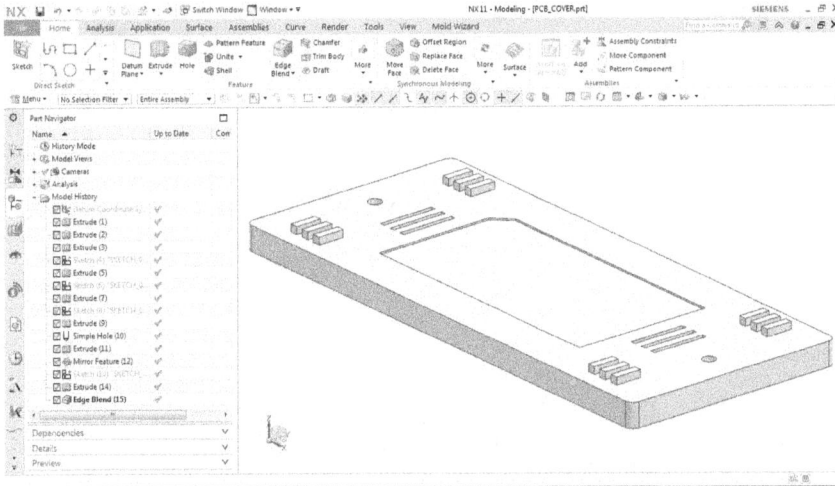

Figure 3-9 The PCB_COVER model in the Modeling environment

Initializing the Project from the Mold Wizard Tab

Now, use the **Initialize Project** tool to define the material and the path for saving the project.

1. Choose the **Mold Wizard** tab.

2. Invoke the **Initialize Project** tool from the **Mold Wizard** tab; the **Initialize Project** dialog box is displayed, as shown in Figure 3-10.

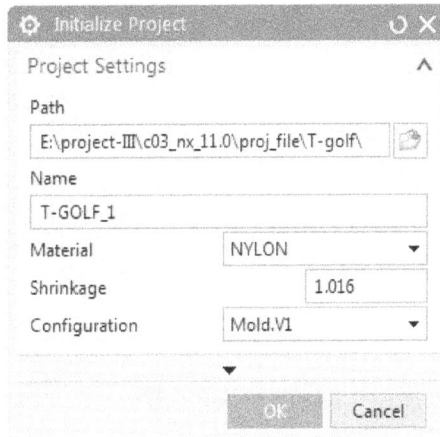

*Figure 3-10 The **Initialize Project** dialog box*

3. Choose the **Browse** button from the **Path** area of the **Project Settings** rollout; the **Open** dialog box is displayed, as shown in Figure 3-11. After setting the path of the project, choose the **OK** button to close the dialog box.

Figure 3-11 The **Open** *dialog box*

4.   Select the **PC+ABS** material from the **Material** drop-down list; the **Shrinkage** edit box gets updated automatically depending upon the type of material selected. Next, select the desired template from the **Configuration** drop-down list. By default, **Mold.V1** template is selected. Choose the **OK** button to close the dialog box; the **New Iray+Ray Traced Studio Rendering** message box is displayed. Choose the **OK** button to close the message box.

Reorienting the Model

Next, you need to use the **Mold CSYS** tool to reorient the model.

1.   Invoke the **Mold CSYS** tool from the **Main** group of the **Mold Wizard** tab; the **Mold CSYS** dialog box is displayed, as shown in Figure 3-12; you are prompted to double-click on WCS.

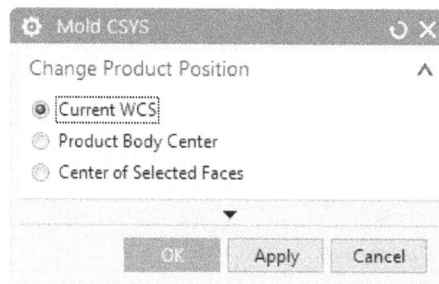

Figure 3-12 The **Mold CSYS** *dialog box*

2.   Double-click on the WCS, refer to Figure 3-13; you are prompted to drag a handle or select a handle. Select the handle of WCS; the **WCS Dynamics** dialog box is displayed.

3.   Select the **XC-YC** rotation handle and then enter **90** in the **Angle** edit box.

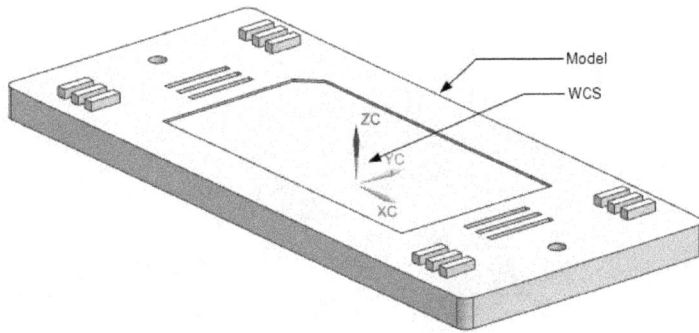

Figure 3-13 *The WCS (Work Coordinate System)*

4. Click the middle mouse button to close the **WCS Dynamics** dialog box.

5. Choose the **OK** button of the **Mold CSYS** dialog box to close the dialog box.

Checking Regions

Now, you need to use the **Check Regions** tool to check the regions of core and cavity.

1. Invoke the **Check Regions** tool from the **Parting Tools** gallery; the **Check Regions** dialog box is displayed, refer to Figure 3-14.

2. Choose the **Calculate** button from the **Calculate** rollout of the **Calculate** tab.

    Next, choose the **Region** tab and then choose the **Set Regions Color** button from the **Define Regions** rollout; the colors of the model get changed. The different regions in the model get updated in different colors representing regions as core, cavity, and undefined regions in the model.

    By default, the **Cavity Region** radio button is selected in the **Assign to Region** rollout.

3. Select the undefined region of the model and then choose the **Apply** button; the undefined region gets the color of the cavity region.

4. Choose the **OK** button to close the dialog box. The modified model after using the **Check Regions** tool is shown in Figure 3-15.

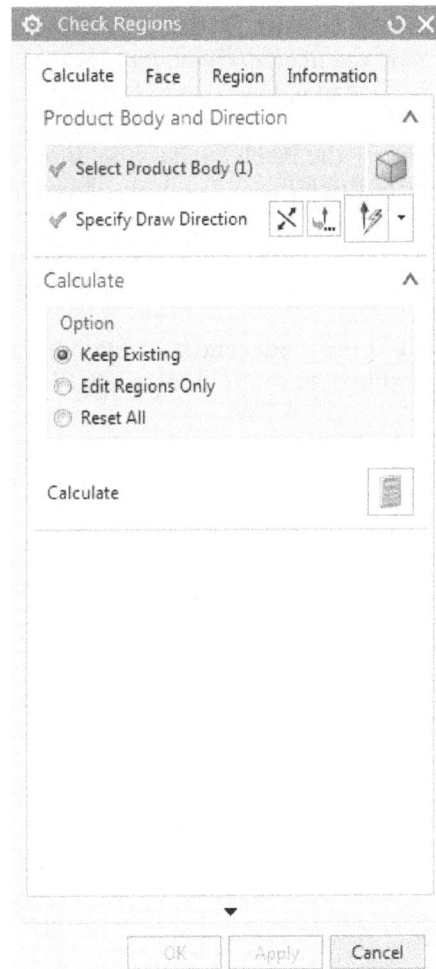

Figure 3-14 *The **Check Regions** dialog box*

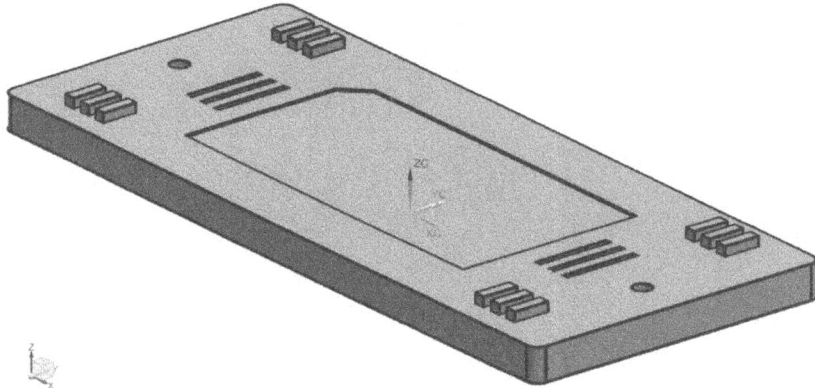

*Figure 3-15 The modified model after using the **Check Regions** tool*

Creating Sheet Bodies to Close the Openings in the Model

Now, you need to create sheet bodies to close the openings in the model by using the **Patch Surface** tool.

1. Choose the **Patch Surface** tool from the **Parting Tools** gallery of the **Mold Wizard** tab; the **Edge Patch** dialog box is displayed, refer to Figure 3-16.

2. Select the **Body** option from the **Type** drop-down list of the **Loop Selection** rollout; you are prompted to select the body.

3. Select the model and then choose the **OK** button to close the dialog box; the model gets modified, as shown in Figure 3-17.

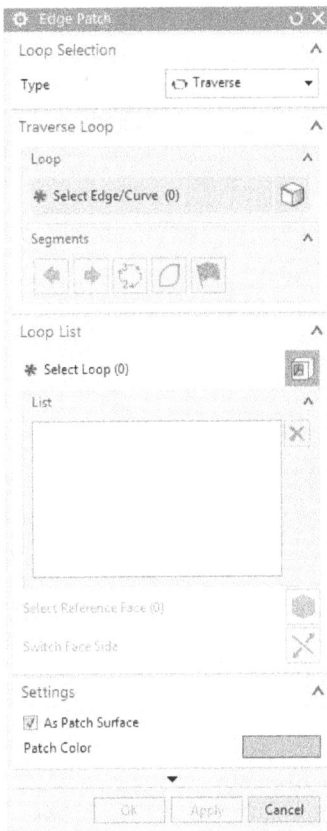

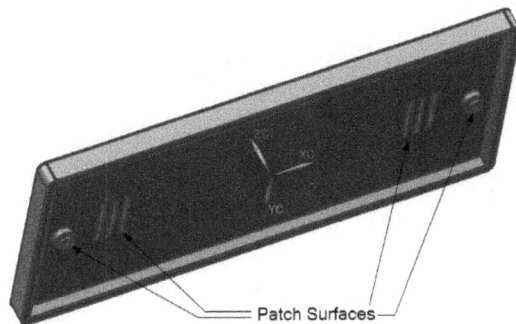

Figure 3-16 The **Edge Patch** *dialog box*   *Figure 3-17* *The model after creating the patch surface*

Creating Core and Cavity Region Sheets

To create parting lines, core, and cavity regions, you need to use the **Define Regions** tool.

1. Choose the **Define Regions** tool from the **Parting Tools** gallery of the **Mold Wizard** tab; the **Define Regions** dialog box is displayed with the **Parting Navigator** window, refer to Figure 3-18.

2. Select **Cavity region** in the **Define Regions** rollout and then select the **Create Regions** and **Create Parting Lines** check boxes from the **Settings** rollout.

3. Choose the **Apply** button; you will notice a green tick mark beside the cavity and core region.

4. Choose the **Cancel** button to close the **Define Regions** dialog box.

5. Uncheck all the check boxes except the **Parting Lines** check box from the **Parting Navigator** window; the parting lines are created in the window.

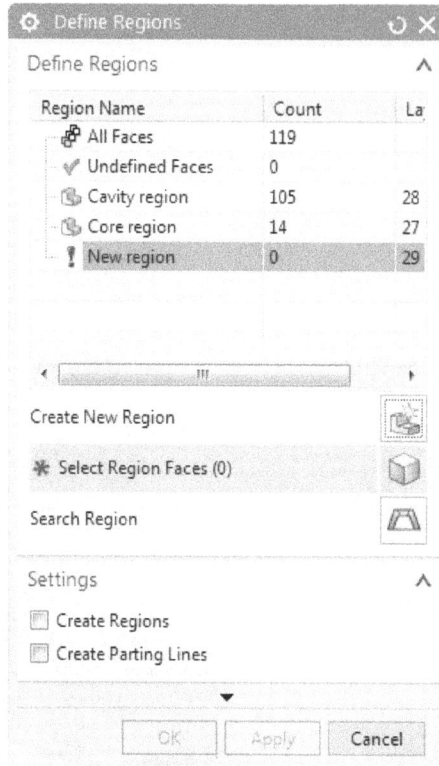

*Figure 3-18 The **Define Regions** dialog box*

Creating a Parting Surface

To create a parting surface, you need to use the **Design Parting Surface** tool.

1. Choose the **Design Parting Surface** tool from the **Parting Tools** gallery of the **Mold Wizard** tab; the **Design Parting Surface** dialog box is displayed, as shown in Figure 3-19.

    By default, the **Bounded Plane** method is selected in the **Create Parting Surface** rollout.

2. Choose the **OK** button; the parting surface of the model is created, as shown in Figure 3-20.

3. Select **PCB_COVER_parting_###** in the assembly navigator and right-click on it; a shortcut menu is displayed. Choose **PCB_COVER_top_###** from the shortcut menu.

4. Choose **File > Save > Save All** to save the file.

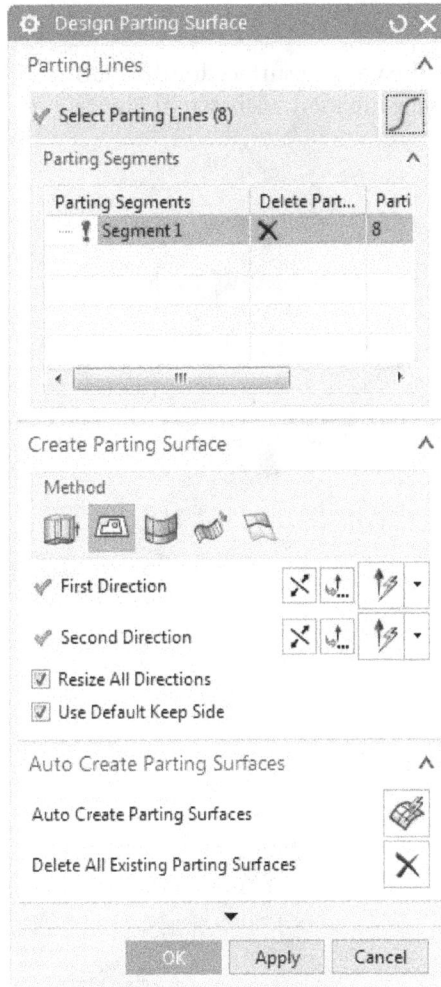

Figure 3-19 *The **Design Parting Surface** dialog box*

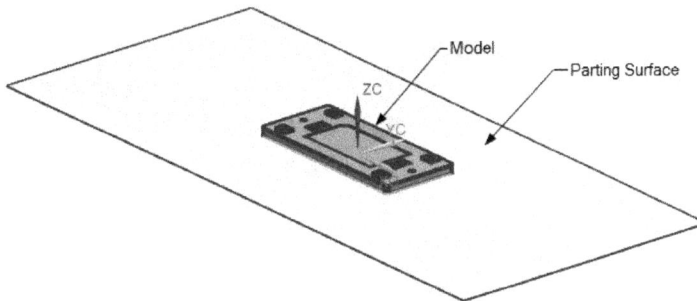

Figure 3-20 *The model with parting surface*

Tutorial 2

In this tutorial, you will create the parting surface for the PCB_UPPER_CASE model contained in Tut 2 folder that you have downloaded as shown in Figure 3-21 and Figure 3-22.

(Expected time: 70 min)

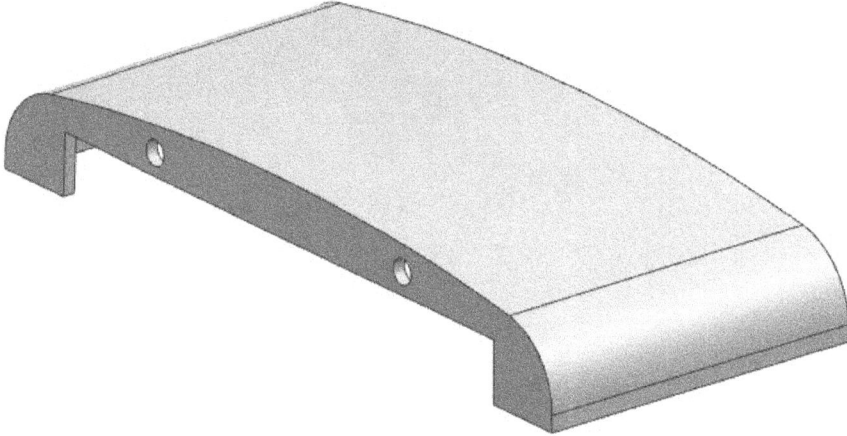

Figure 3-21 The PCB_UPPER_CASE model

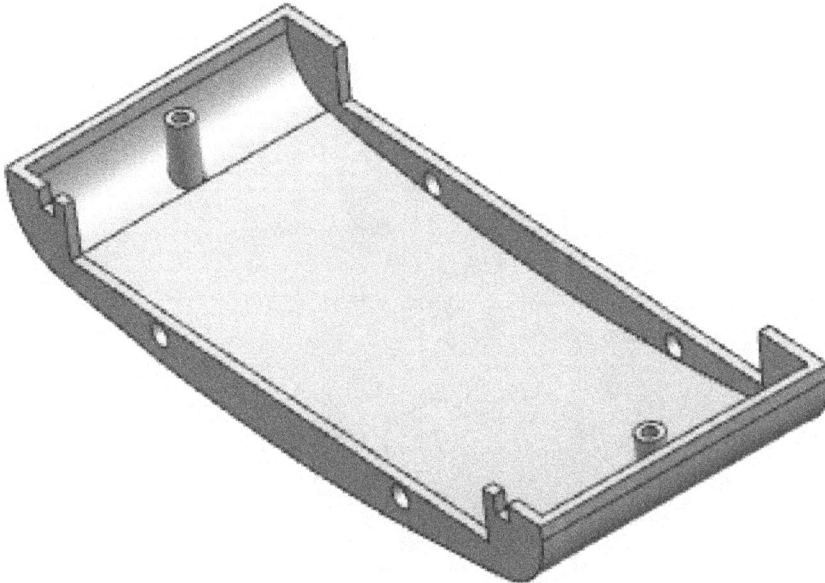

Figure 3-22 The other view of the PCB_UPPER_CASE model

The following steps are required to complete this tutorial:

a.   Open the model.
b.   Initialize the project.
c.   Reorient the model by using the **Mold CSYS** tool.
d.   Analyze the core and cavity regions by using the **Check Regions** tool.
e.   Patch the surface by using the **Patch Surface** tool.
f.   Specify the core and cavity region by using the **Define Regions** tool.
g.   Create a parting surface by using the **Design Parting Surface** tool.
h.   Save the model.

Starting NX and Opening the Model

First, you need to start NX and then open a new file.

1.   Double-click on the shortcut icon of NX available on the desktop of your computer to start NX.

2.   Choose the **Open** button from the **Standard** group of the **Home** tab or choose **Menu > File > Open** from the **Top Border Bar**; the **Open** dialog box is displayed.

3.   In this dialog box, select **PCB_UPPER_CASE** from the **Name** list; **PCB_UPPER_CASE** is displayed in the **File name** drop-down list, refer to Figure 3-23. Then, choose the **OK** button; the model is displayed in the modeling environment, refer to Figure 3-24.

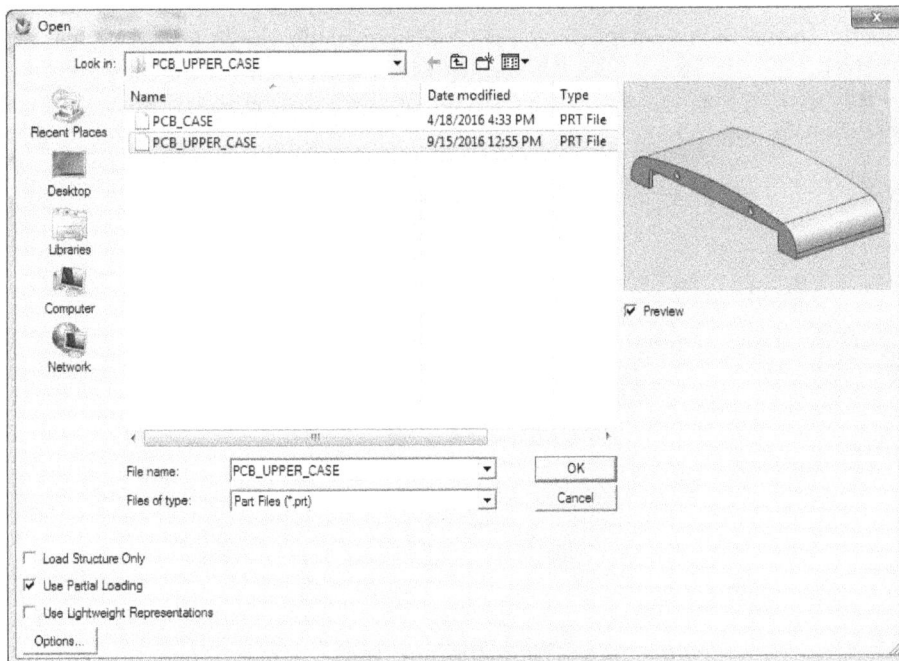

Figure 3-23 The **Open** dialog box

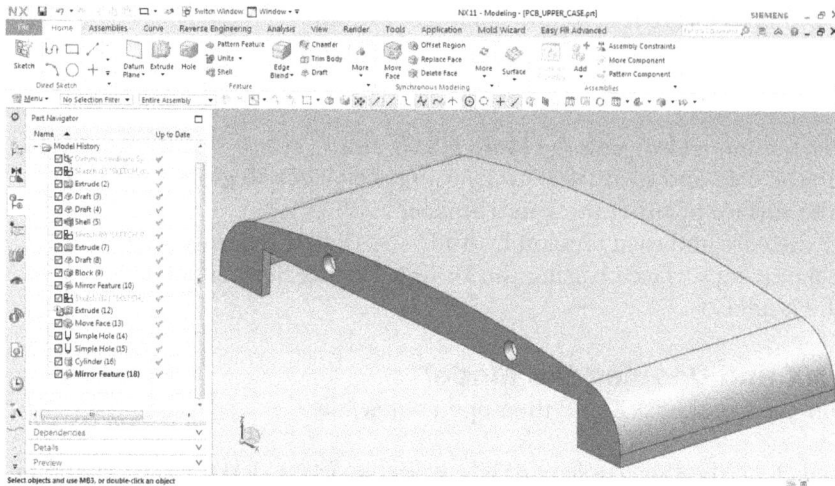

Figure 3-24 The PCB_UPPER_CASE model in the Modeling environment

Initializing the Project from the Mold Wizard Tab

Now, you need to **Initialize Project** tool to define the material and the path for saving the project.

1. Choose the **Mold Wizard** tab.

2. Choose the **Initialize Project** tool from the **Mold Wizard** tab; the **Initialize Project** dialog box is displayed, refer to Figure 3-25. Choose the **OK** button from the dialog box and the **New Iray+Ray Traced Studio Rendering** message box gets displayed, as shown in Figure 3-26. Choose the **OK** button to close the message box.

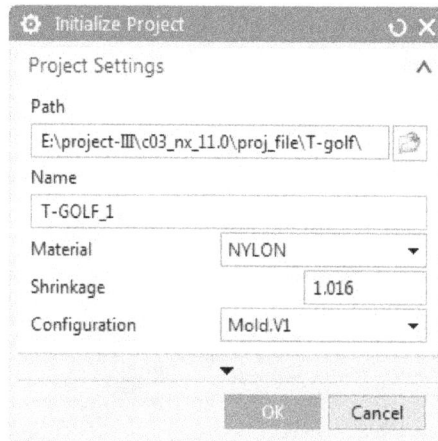

Figure 3-25 The **Initialize Project** dialog box

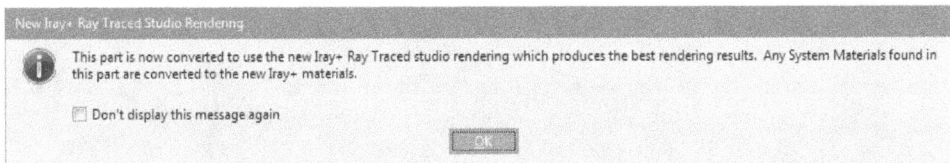

Figure 3-26 The New Iray+Ray Traced Studio Rendering message box

3. Choose the **Browse** button from the **Path** area of the **Project Settings** rollout; the **Open** dialog box will be displayed where you can set the path for the project. Select a path for the project and then choose the **OK** button.

4. Select the **PC+ABS** material from the **Material** drop-down list; the **Shrinkage** edit box gets updated automatically depending upon the type of selection of material. Next, select the desired template from the **Configuration** drop-down list. By default, the **Mold.V1** template is selected. Choose the **OK** button to close the dialog box; the **New Iray+Ray Traced Studio Rendering** message box is displayed again. Choose the **OK** button to close the message box.

Reorienting the Model

In this section, you need to reorient the model by using the **Mold CSYS** tool.

1. Choose the **Mold CSYS** tool from the **Main** gallery of the **Mold Wizard** tab; the **Mold CSYS** dialog box is displayed, as shown in Figure 3-27, and you are prompted to double-click on WCS.

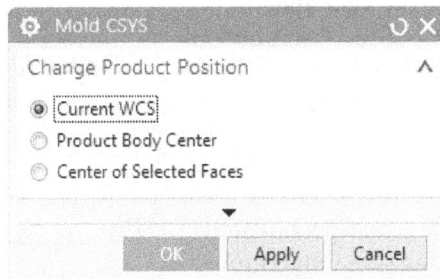

Figure 3-27 The Mold CSYS dialog box

2. Double-click on WCS; the **WCS Dynamics** dialog box is displayed, refer to Figure 3-28.

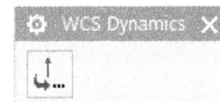

Figure 3-28 The WCS Dynamics dialog box

3. Select the **XC-YC** rotation handle and then enter **90** in the **Angle** edit box and press Enter.

4. Click the middle mouse button to close the **WCS Dynamics** dialog box. Next, choose the **OK** button from the **Mold CSYS** dialog box; you will notice that the orientation of the model gets changed, refer to Figure 3-29.

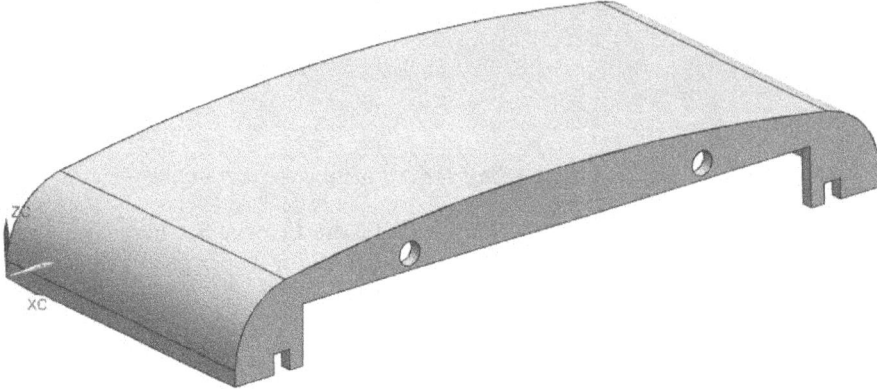

Figure 3-29 The changed orientation of the model

Checking Regions

In this section, you need to use the **Check Regions** tool to check regions of core and cavity.

1. Invoke the **Check Regions** tool from the **Parting Tools** gallery; the **Check Regions** dialog box will be displayed, refer to Figure 3-30.

2. In this dialog box, choose the **Calculate** button from the **Calculate** rollout of the **Calculate** tab and then choose the **Region** tab. Next, choose the **Set Regions Color** button from the **Define Regions** rollout; the colors of the model get changed.

    The different regions in the model get updated in different colors representing different regions as core, cavity, and undefined regions in the model. The **Core Region** radio buttton is selected by default in the **Assign to Region** rollout. Now, select the undefined region of the model and then choose the **Apply** button. The undefined region gets the color of the core region.

3. Choose the **OK** button to close the dialog box. The modified model after using the **Check Region** tool is shown in Figure 3-31.

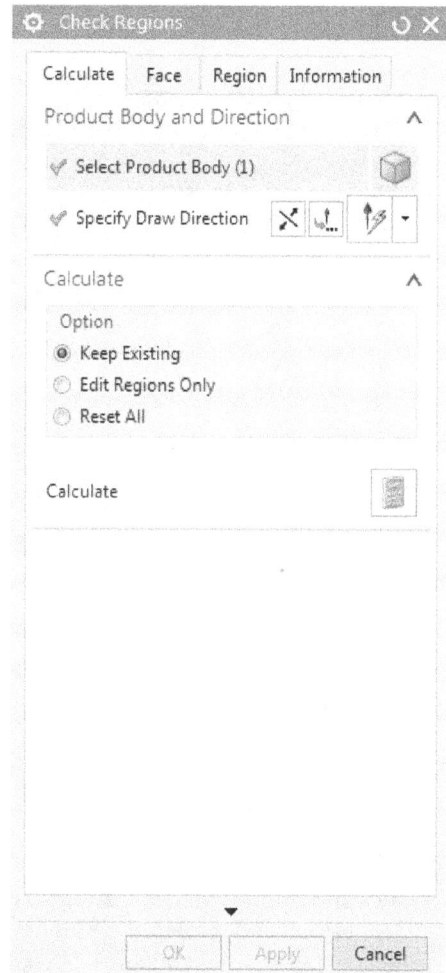

Figure 3-30 The Check Regions dialog box

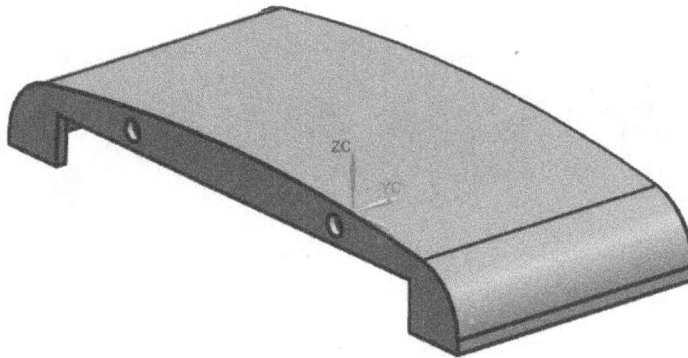

Figure 3-31 *The modified model after using the **Check Region** tool*

Creating Sheet Bodies to Close the Openings in the Model

Now, you need to create sheet bodies to close the openings in the model by using the **Patch Surface** tool.

1. Choose the **Patch Surface** tool from the **Parting Tools** gallery of the **Mold Wizard** tab; the **Edge Patch** dialog box is displayed, refer to Figure 3-32.

    By default, the **Transverse** option is selected from the **Type** drop-down list of the **Loop Selection** rollout and you are prompted to select the edges or curves.

2. Select the edge of hole in the model. You will notice green tick mark in the **Select Loop** and **Select Reference Face** areas of the **Loop List** rollout and then choose the **Edge** or **Curve** button to add another loop to the **List** sub-rollout of the **Loop List** rollout. Continue patching the holes until the four holes are selected in the model.

3. Select **Loop1** from the **List** sub-rollout and then choose the **Apply** button. You will notice that the hole is filled by a surface. Repeat until the four holes are filled.

4. Choose the **Cancel** button to close the dialog box. Final model after patching is shown in Figure 3-33.

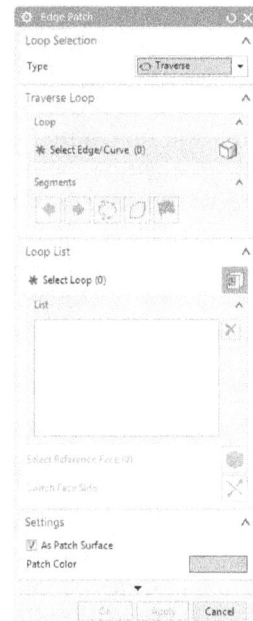

Figure 3-32 *The **Edge Patch** dialog box*

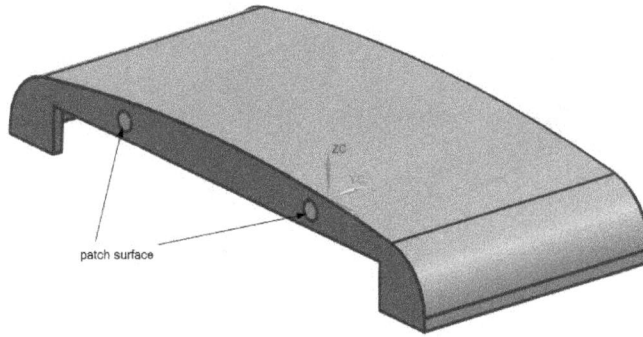

Figure 3-33 *The model after using the* *Patch Surface* *tool*

Creating Core and Cavity Region Sheets

To create parting lines, core, and cavity regions, you need to use the **Define Regions** tool.

1. Choose the **Define Regions** tool from the **Parting Tools** gallery of the **Mold Wizard** tab; the **Define Regions** dialog box is displayed, refer to Figure 3-34.

    You will notice that the total faces in the model are 34 which can be seen in the **Define Regions** rollout. Out of the 34 faces, 7 are in the cavity region, 23 are in the core region, and 4 are undefined faces.

2. Select the **Cavity region** in the **Define Regions** rollout and then select the **Create Regions** and **Create Parting Lines** check boxes from the **Settings** rollout.

3. Choose the **Apply** button. You will notice the green tick mark beside the **Cavity region** and the **Core region** in the **Define Regions** rollout.

4. Choose the **Cancel** button to close the **Define Regions** dialog box.

5. Choose **Parting Navigator** tool from the **Parting Tools** gallery of the **Mold Wizard** tab; the **Parting Navigator** tree structure is displayed, refer to Figure 3-35. Make sure only the **Parting Lines** check box is selected in the **Parting Navigator** tree structure. If other check boxes are selected then clear them.

    Figure 3-36 shows the parting lines of the model.

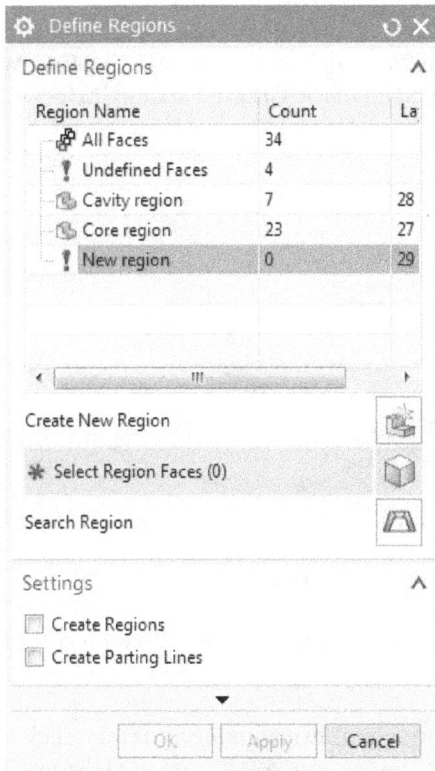

Figure 3-34 The **Define Regions** dialog box

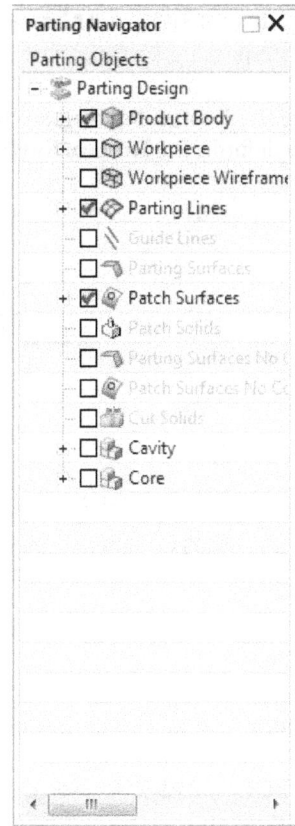

Figure 3-35 The **Parting Navigator** tree structure

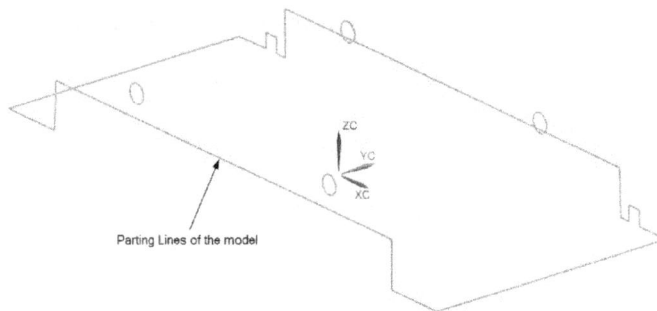

Figure 3-36 The parting lines of the model

Creating a Parting Surface

To create parting surface, you need to use the **Design Parting Surface** tool.

1. Choose the **Design Parting Surface** tool from the **Parting Tools** gallery of the **Mold Wizard** tab. By default, the **Ribbon Surface** method is selected in the **Create Parting Surface** rollout.

2. Choose the **OK** button to create the parting surface of the model, refer to Figure 3-37.

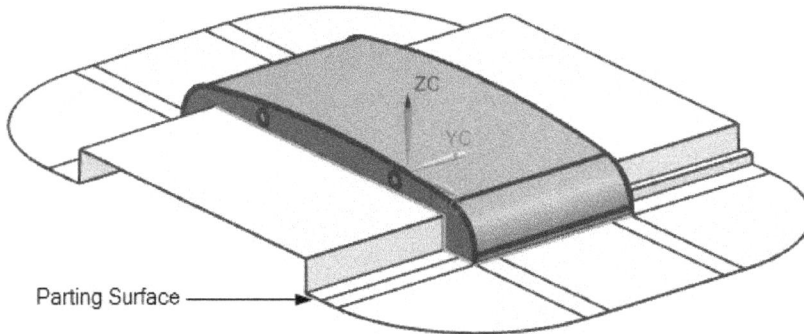

Figure 3-37 *The parting surface of the model*

Note that, you need to select the check box in corresponding to **Product Body** in the **Parting Navigator** tree to see the model.

3. Select **PCB_UPPER_CASE_parting_###** in the assembly navigator and right-click on it; a shortcut menu is displayed. Select **PCB_UPPER_CASE_top_###** from the **Display Parent** shortcut menu.

Saving and Closing the File

1. Choose **Menu > File > Close > Save and Close** from the **Top Border Bar** to save and close the file.

Self-Evaluation Test

Answer the following questions and then compare them to those given at the end of this chapter:

1. The _____ tool is used to define path of the mold project.

2. Parting Surfaces are classified as _____ and _____ parting surfaces.

3. There are _____ types of the **Configuration** in the **Initialize Project** dialog box.

4. You can change the orientation of the model by choosing the _____ tool.

5. The _____ tool is used to apply material to the model.

6. In the **Mold CSYS** dialog box, _____ methods are available to reorient the model.

7. The _____ tool is used to assign core and cavity regions to the model.

8. The _____ tool is used to check the undercut of the model.

9. You can split the face of a model. (T/F)

10. The **Patch Surface** tool is used to create the sheet to close the openings in a model. (T/F)

Review Questions

Answer the following questions:

1. Which of the following dialog boxes is displayed when you choose the **Patch Surface** tool?

    (a) **Patch Surface**   (b) **Edge Patch**
    (c) **Patch**   (d) None of these

2. Which of the following tools in NX Mold Wizard is used to create core and cavity regions?

    (a) **Check Regions**   (b) **Define Regions**
    (c) **Patch Surface**   (d) None of these

3. The **Patch Surface** tool has _____ options to patch the model.

4. The _____ tool is used to create parting surface of the model.

5. The _____ tool is used to check the draft angle of the model.

6. The **Transverse** option is used to patch the surface. (T/F)

7. The **Face** option is used to patch the surface. (T/F)

8. The **Ribbon Surface** option is used to create the parting surface. (T/F)

9. You can edit material database in NX Mold Wizard. (T/F)

10. You can select the types of template such as **Mold.V1**, **ESI**, and **Original** from the **Configuration** drop-down list. (T/F)

EXERCISES

To perform the exercises, you need to download the zipped file named as *c03_NX_Mold_input* from the Input Files section of the CADCIM website. The complete path for downloading the file is:

> *Textbooks > CAD/CAM > NX_Mold > Mold Design using NX 11.0: A Tutorial Approach > Input Files*

After the file is downloaded, extract the folder.

Exercise 1

In this exercise, you will open the exr_01 file. Create the parting surface of the model, as shown in Figure 3-38 and save it.   (**Expected time: 60 min**)

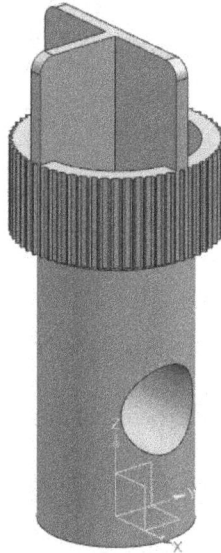

Figure 3-38 Model for Exercise 1

Exercise 2

In this exercise, you will open the exr_02 file. Create the parting surface of the model, as shown in Figure 3-39 and save it.   (**Expected time: 60 min**)

Figure 3-39 Model for Exercise 2

Answers to Self-Evaluation Test

1. Initialize Project, **2.** flat and non-flat, **3.** three, **4. Mold CSYS**, **5. Initialize Project**, **6.** three, **7. Check Regions**, **8. Check Regions**, **9.** T, **10.** T

Chapter 4

Creating Core and Cavity

Learning Objectives

After completing this chapter, you will be able to:
- *Define workpiece*
- *Create cavity layout*
- *Define cavity and core*

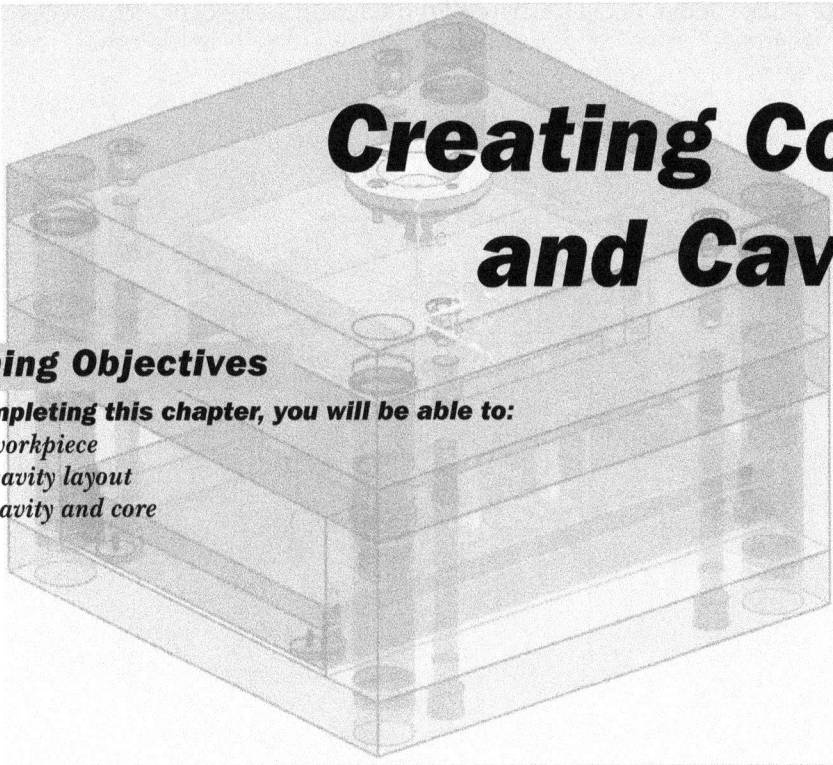

INTRODUCTION

In the previous chapter, you learned creating a parting surface. Once the parting surface has been created, you need to create core, cavity and layout for core and cavity. In this chapter, you will study how you can create core and cavity using various tools.

CORE AND CAVITY

The core and cavity of a component are generally considered male and female portions of the component. Generally, core provides inner profile of the component and cavity provides outer profile of the component. The shape of the core and cavity depends on the profile of the component. Figure 4-1 shows the core and cavity. When these two halves come together, a space is formed between core and cavity which is the shape of the component.

Now, you will learn how to create multiple cores and cavities.

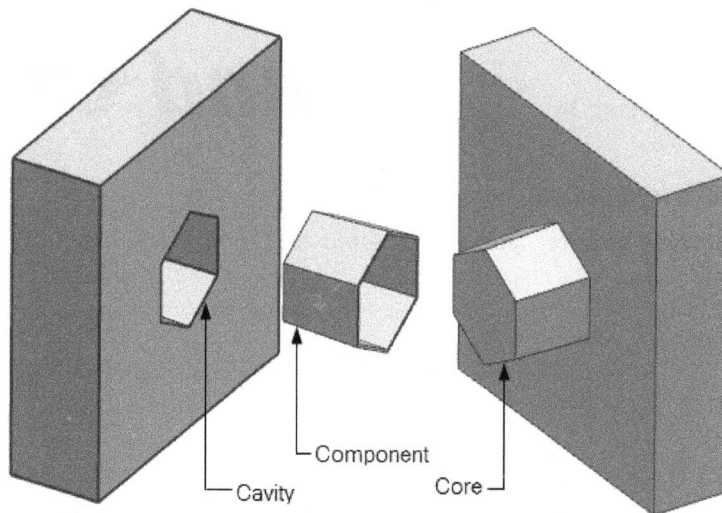

Figure 4-1 *Basic mold consisting of core and cavity*

Note
*Use the **Initialize Project** tool before using the tools explained in this chapter.*

Workpiece

Ribbon: Mold Wizard > Main > Workpiece

This tool helps you define the size of core and cavity. To invoke this tool, choose the **Workpiece** tool from the **Main** gallery of the **Mold Wizard** tab; the **New Iray + Ray Traced Studio Rendering** window is displayed. Choose the **OK** button; the **Workpiece** dialog box is displayed, refer to Figure 4-2. The options in this dialog box are discussed next.

Type Rollout

The options in the drop-down list available in this rollout are used to create core and cavity for single or multiple components.

Product Workpiece

This option helps you create core and cavity only for single component.

Combined Workpiece

This option helps you create the core and cavity for multiple components.

Workpiece Method Rollout

The options in the **Workpiece Method** drop-down list are used to define the way you want to create the workpiece.

The options available in the **Workpiece Method** drop-down list are:

(i)   User Defined Block
(ii)   Cavity-Core
(iii)   Cavity Only
(iv)   Core Only

Dimensions Rollout

The options in this rollout helps you to define the size of the workpiece. After specifying the size of the workpiece, choose the **OK** button to close the dialog box; the **Information** window is displayed. Choose the close button to close the dialog box.

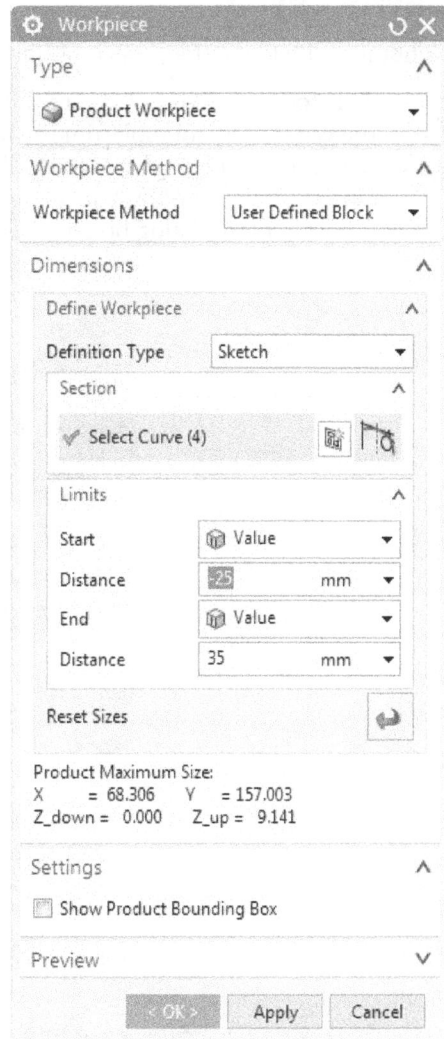

Figure 4-2   The **Workpiece** dialog box

Cavity Layout

Ribbon: Mold Wizard > Main > Cavity Layout

This tool helps you to create the layout of core and cavity workpiece. To do so, choose the **Cavity Layout** tool from the **Main** gallery of the **Mold Wizard** tab; the **Cavity Layout** dialog box will be displayed, refer to Figure 4-3. Also, the workpiece will be selected automatically, as shown in Figure 4-4. The options available in various rollouts of the **Cavity Layout** dialog box are discussed next.

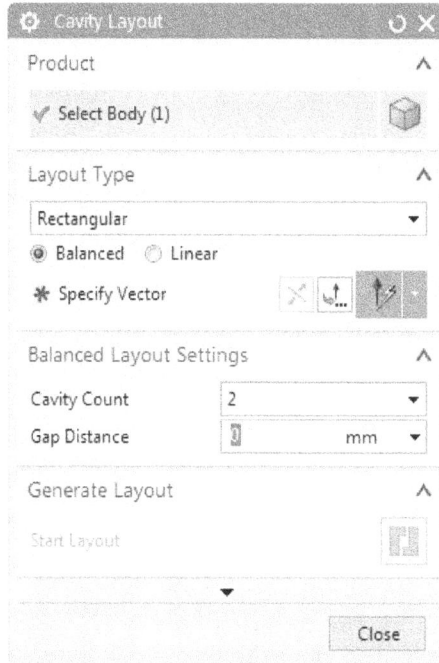

*Figure 4-3 The **Cavity Layout** dialog box*

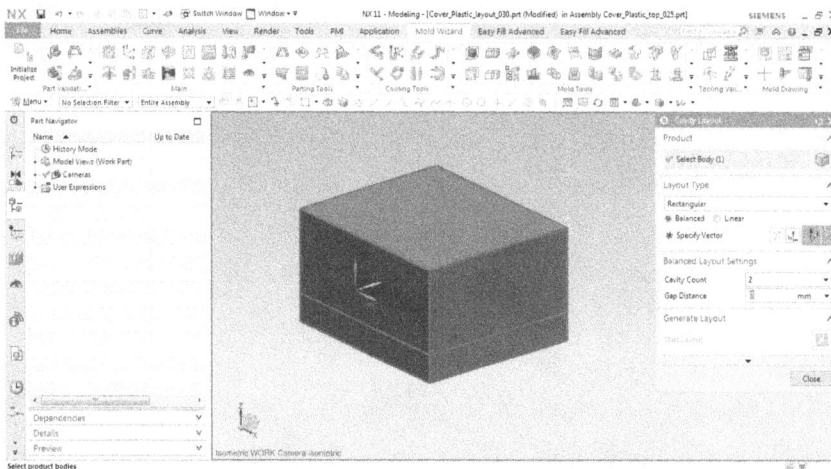

Figure 4-4 The workpiece selected automatically

Product Rollout
This rollout is used to select the workpiece.

Layout Type Rollout
The options in this rollout are used to specify the type of layout you need to create. The options in this rollout of the **Cavity Layout** dialog box are discussed next.

Layout Type Drop-down
The options in this drop-down list are used to define the type of layout. The **Rectangular** option in this drop-down list is selected by default and helps you to create rectangular layout of the workpiece. You can also choose the **Circular** option from this drop-down list to create a circular layout. The **Specify Vector** area in the **Layout Type** rollout helps you to define the direction placement of the component with respect to other component. Specify the parameter of the layout in the **Balanced Layout Settings** or **Circular Layout Settings** rollout depending upon the option selected in the drop-down list in the Layout Type rollout. Choose the **Start Layout** button from the **Generate Layout** rollout to create rectangular or circular layout; preview of the layout will be displayed in the window. After that, choose the **Edit Insert Pocket** button from the **Edit Layout** rollout; the **Insert Pocket** dialog box will be displayed, refer to Figure 4-5. Select suitable

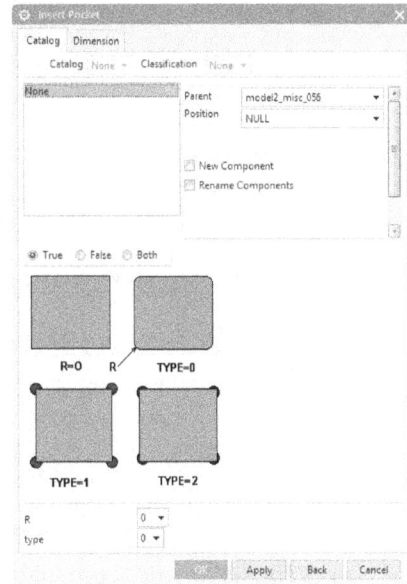

*Figure 4-5 The **Insert Pocket** dialog box*

options from the **R** and **type** drop-down lists. Choose the **OK** button and then the **Close** button from the **Cavity Layout** dialog box. Figure 4-6 and 4-7 show the rectangular and circular layout respectively.

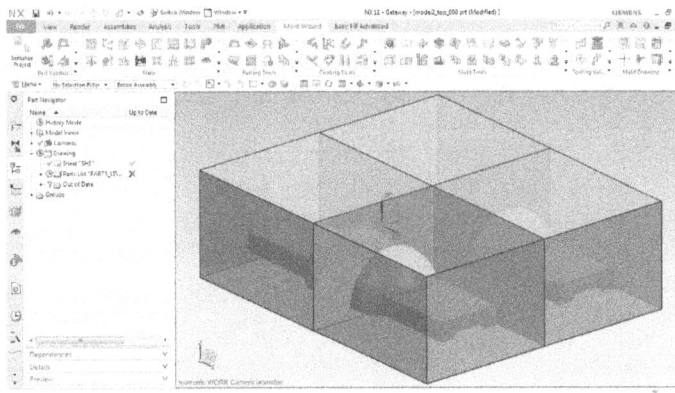

Figure 4-6 The rectangular layout of workpiece

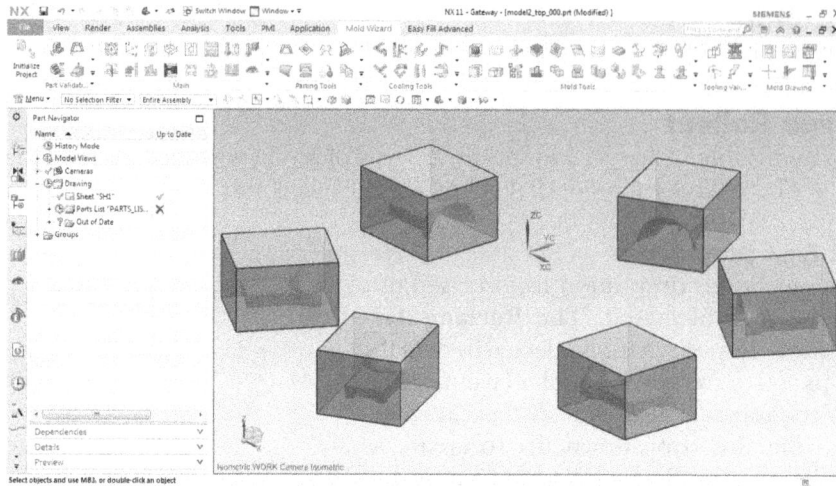

Figure 4-7 *The circular layout of workpiece*

Define Cavity and Core

Ribbon: Mold Wizard > Parting Tools > Define Cavity and Core

This tool helps you to create core and cavity of the component. To create core and cavity of the model, invoke the **Define Cavity and Core** tool from the **Parting Tools** gallery of the **Mold Wizard** tab; the **Define Cavity and Core** dialog box will be displayed, refer to Figure 4-8. The **Cavity region** area is selected by default in the **Select Sheets** rollout of the dialog box. Choose the **Apply** button; the **New Iray + Ray Traced** Studio Rendering window will be displayed. Choose the **OK** button; the window will be closed and the **View Parting Result** window will be displayed along with the **Information** window. Ensure that the cavity is created otherwise choose the **Reverse Normal** button. Choose the **OK** button to close the window; the **Define Cavity and Core** dialog box will be displayed again. Select the **Core region** area from the **Select Sheets** rollout. Choose the **Apply** button; the **New Iray + Ray Traced Studio Rendering** window will be displayed. Choose the **OK** button; the **View Parting Result** window will be displayed. In case, the core is not created, choose the **Reverse Normal** button. Choose the **OK** button; the window will be closed and the **Define Cavity and Core** dialog box gets displayed again. Choose the **Cancel** button to close the dialog box.

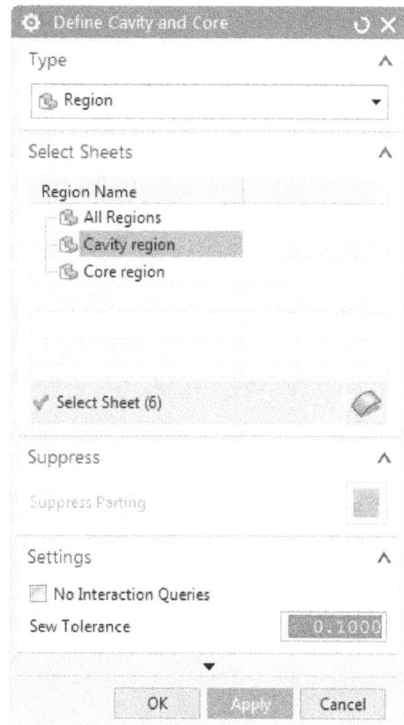

Figure 4-8 *The **Define Cavity and Core** dialog box*

TUTORIALS

To perform the tutorials, you need to download the zipped file named as *c04_NX_Mold_input* from the **Input Files** section of the CADCIM website. The complete path for downloading the file is:

> *Textbooks > CAD/CAM > NX_Mold > Mold Design using NX 11.0: A Tutorial Approach >*
> *Input Files*

After the file is downloaded, extract the folder. In this folder, you will find Tut1 and Tut2 folders containing input files for Tutorial 1 and Tutorial 2.

Tutorial 1

In this tutorial, you will create core, cavity and the layout of core and cavity to the model (PCB_COVER_top_###) contained in Tut 1 folder that you have downloaded. Refer to Figure 4-9 for model.   **(Expected time: 30 min)**

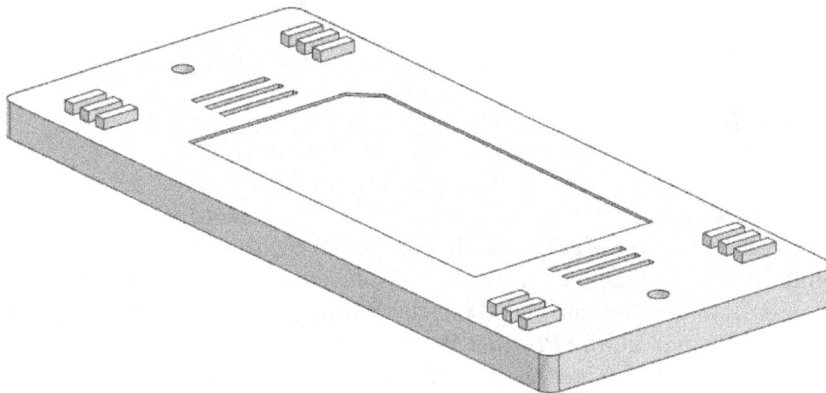

Figure 4-9 The PCB_COVER model for Tutorial 1

The following steps are required to complete this tutorial:

a. Open the model.
b. Define insert size by using the **Workpiece** tool.
c. Define type of layout and number of cavities.
d. Create core and cavity using the **Define Cavity and Core** tool.
e. Save the model.

Starting NX and Opening a Model

First, you need to start NX and then open a file.

1. Double-click on the shortcut icon of NX on the desktop of your computer to start NX.

2. Choose the **Open** button from the **Standard** group of the **Home** tab or choose **Menu > File > Open** from the **Top Border Bar**; the **Open** dialog box is displayed.

3.   Browse to **PCB_COVER_top_###** in Tut 1 folder; the **PCB_COVER_top_###** is displayed
     in the **File name** drop-down list. Next, choose the **OK** button; the model is displayed, refer
     to Figure 4-10.

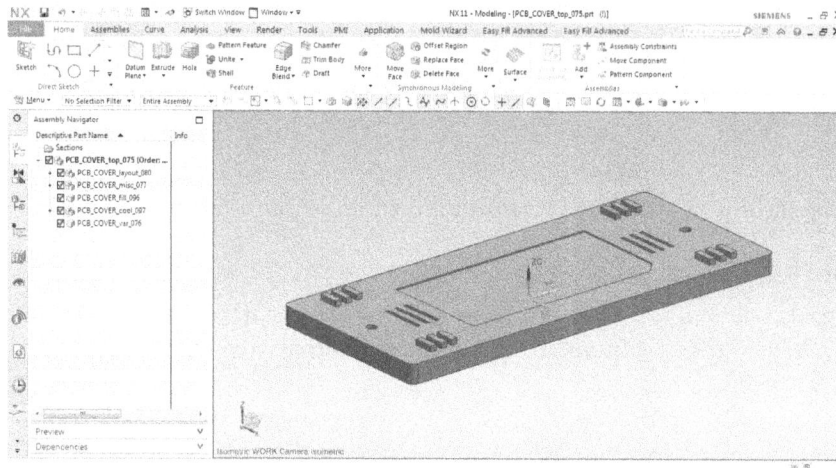

Figure 4-10 *The PCB_COVER model in modeling environment*

Creating Workpiece

You need to create the workpiece by using the **Workpiece** tool.

1.   Choose the **Workpiece** tool from the **Main** gallery of the **Mold Wizard** tab; the **New Iray
     + Ray Traced Studio Rendering** window is displayed, refer to Figure 4-11. Choose the
     OK button; the **Workpiece** message box is displayed, refer to Figure 4-12(a). Choose the
     OK button button from the message box; the **Workpiece** dialog box is displayed, refer to
     Figure 4-12(b). By default, the **Product Workpiece** option is selected in the drop-down list
     of the **Type** rollout and the **User Defined Block** option is selected in the **Workpiece Method**
     drop-down list. Refer to Figure 4-13 for the size of the core and cavity insert. Choose the
     OK button to close the dialog box.

Note the size of insert is automatically defined.

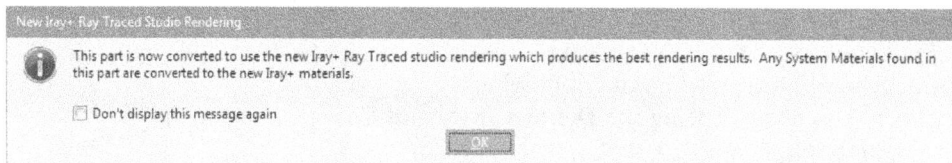

Figure 4-11 *The **New Iray + Ray Traced Studio Rendering** window*

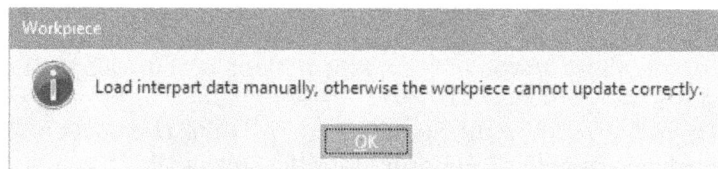

Figure 4-12(a) *The **Workpiece** message box*

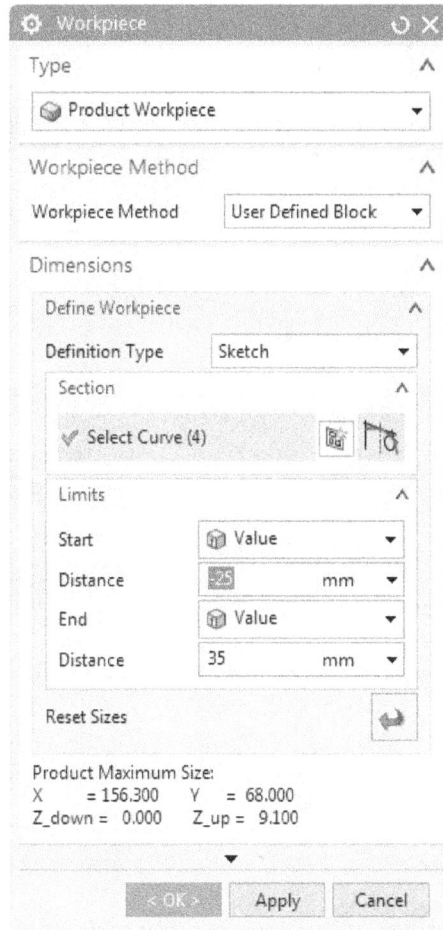

Figure 4-12(b) The **Workpiece** dialog box

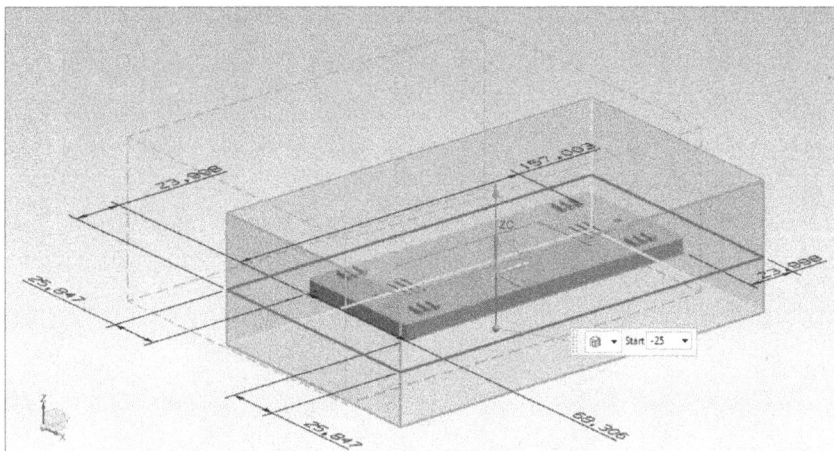

Figure 4-13 The core and cavity insert

Creating Workpiece Layout

In this section, you will use the **Cavity Layout** tool to create the layout of the core and cavity insert.

1. Choose the **Cavity Layout** tool from the **Main** gallery of the **Mold Wizard** tab; the **Cavity Layout** dialog box is displayed, refer to Figure 4-14. Also the body of the core and cavity automatically gets selected and a green tick mark becomes visible in the **Select Body** area of the **Product** rollout. By default, the **Rectangular** option is selected in the **Layout Type** rollout.

   By default, the **Balanced** radio button is selected in the **Layout Type** rollout.

2. Select **4** from drop-down list in the **Cavity Count** edit box of the **Balanced Layout Settings** rollout.

3. Click in the **Specify Vector** area of the **Layout Type** rollout and specify the **XC** axis as a direction. Choose the **Start Layout** button from the **Generate Layout** rollout; the layout gets created, refer to Figure 4-15.

4. Choose the **Auto Center** button from the **Edit Layout** rollout to align the center of layout with the center of the mold base.

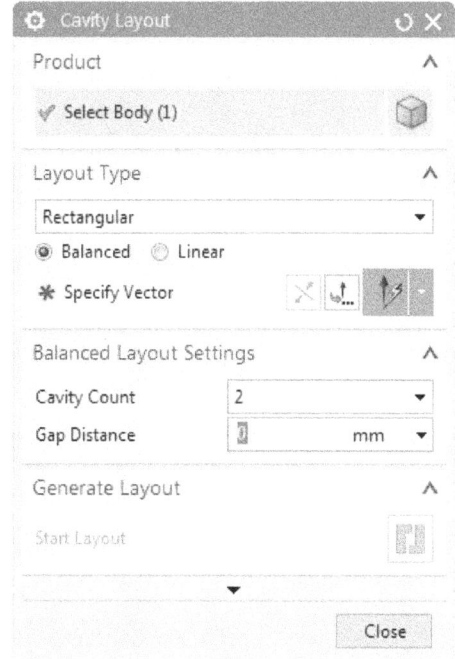

Figure 4-14 The Cavity Layout dialog box

5. Choose the **Edit Insert Pocket** button from the **Edit Layout** rollout; the **Insert Pocket** dialog box is displayed, refer to Figure 4-16.

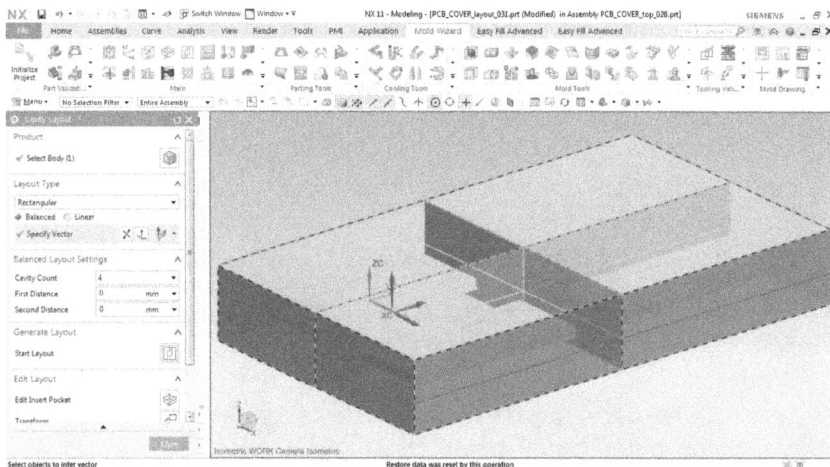

Figure 4-15 The cavity layout of the workpiece

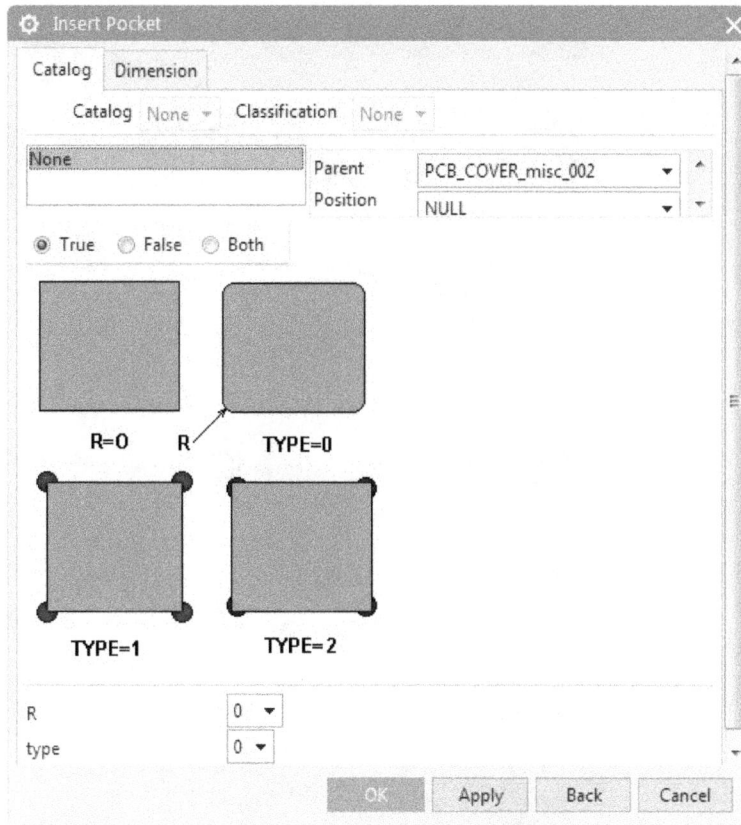

*Figure 4-16 The **Insert Pocket** dialog box*

6. Select **5** from the **R** drop-down list and **2** from the **type** drop-down list. Next, choose the **OK** button to close the dialog box; the **Cavity Layout** dialog box reappear.

7. Choose the **Close** button to close the **Cavity Layout** dialog box.

Creating Core and Cavity

Now, you will create core and cavity by using the **Define Cavity and Core** tool.

1. Choose the **Define Cavity and Core** tool from the **Parting Tools** gallery of the **Mold Wizard** tab; the **Define Cavity and Core** dialog box is displayed, refer to Figure 4-17.

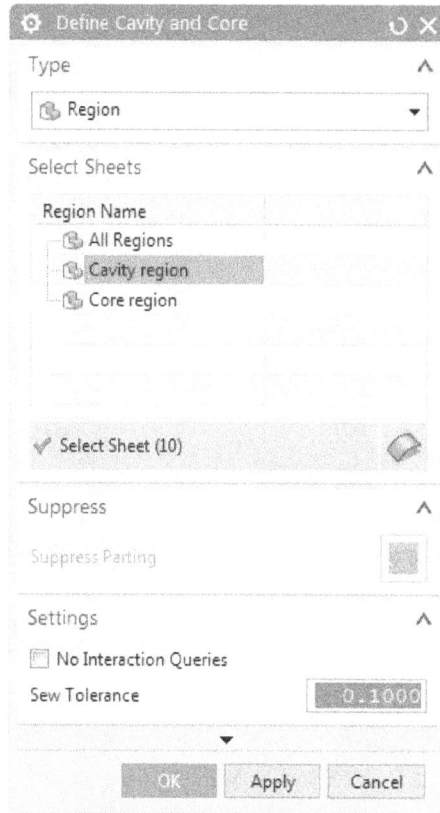

Figure 4-17 The *Define Cavity and Core* dialog box

By default, the **Region** option is selected in the **Type** drop-down list of the **Type** rollout and the **Cavity region** is highlighted in the **Select Sheets** rollout.

2. Choose the **Apply** button, the **New Iray + Ray Traced Studio Rendering** window is displayed. Choose the **OK** button; the **View Parting Result** dialog box and the cavity of the model get displayed. Refer to Figure 4-18 for cavity of the model. Choose the **OK** button; the dialog box is closed and a green tick mark is displayed before the **Cavity region** in the **Select Sheets** rollout of the dialog box. Now, select the **Core region** in the **Select Sheets** rollout of the dialog box and repeat the same procedure as for cavity; the core is displayed, refer to Figure 4-19.

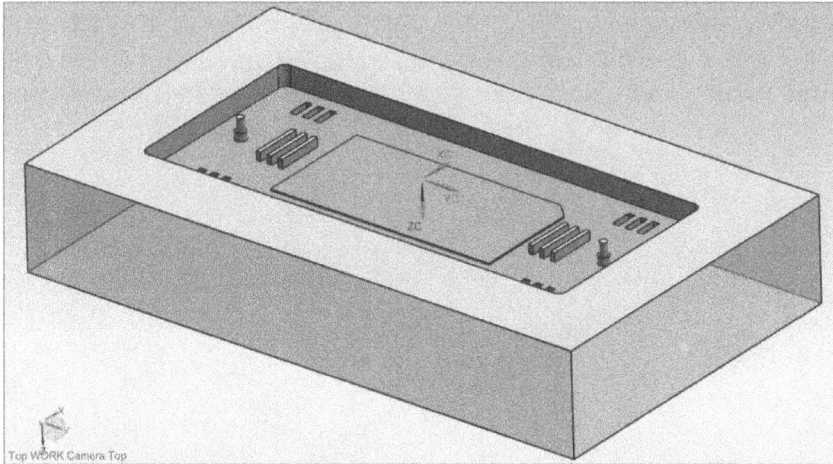

Figure 4-18 *The resultant cavity of the component*

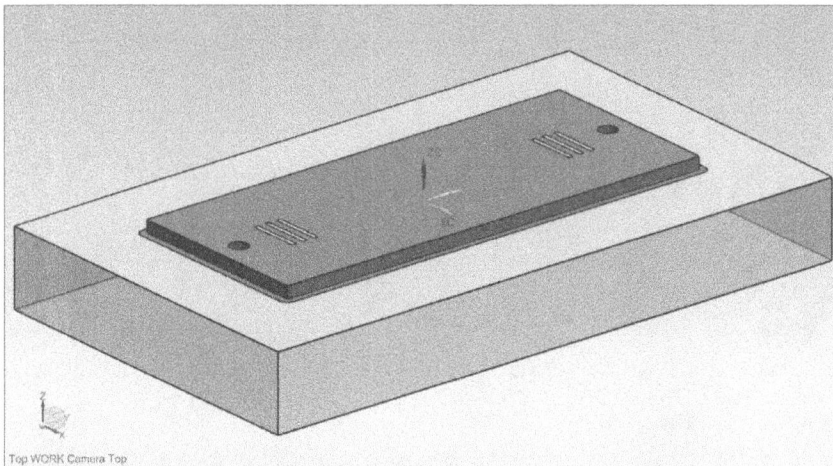

Figure 4-19 *The resultant core of the component*

3. Next, choose the **Cancel** button to close the dialog box.

4. Now, select **PCB_COVER_parting_###** in the assembly navigator and right-click on it; a shortcut menu is displayed. Choose **PCB_COVER_top_###** from the **Display Parent** cascading menu.

Saving and Closing the File

1. Choose **Menu > File > Close > Save and Close** from the **Top Border Bar** to save and close the file.

Tutorial 2

In this tutorial, you will create the core, cavity and the core and cavity layout to the model (PCB_UPPER_CASE_top_###) contained in Tut 2 folder that you have downloaded. Refer to Figure 4-20 and Figure 4-21 for the model.   **(Expected time: 30 min)**

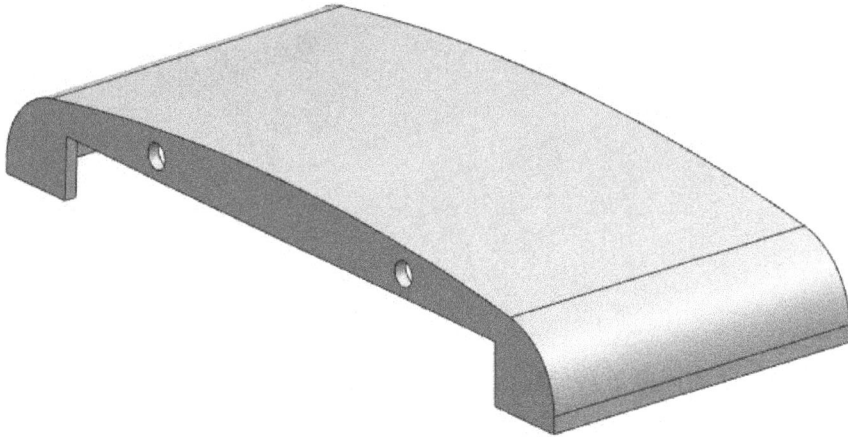

Figure 4-20 *The PCB_UPPER_CASE model*

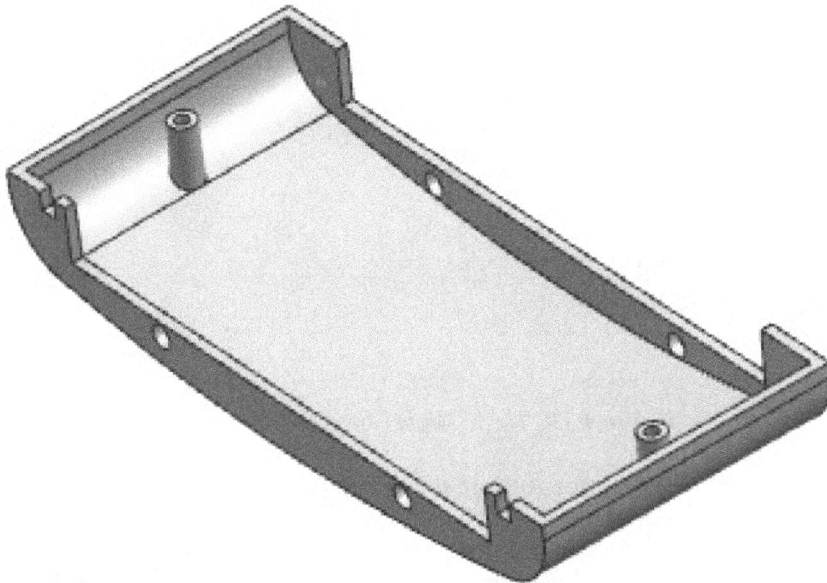

Figure 4-21 *The other view of the PCB_UPPER_CASE model*

The following steps are required to complete this tutorial:

a.   Open the model.
b.   Define insert size by using the **Workpiece** tool.
c.   Define type of layout and number of cavities.
d.   Create core and cavity using the **Define Cavity and Core** tool.
e.   Save the model.

Starting NX and Opening the Model

First, you need to start NX and then open a file.

1.   Double-click on the shortcut icon of NX on the desktop of your computer to start NX.

2.   Choose the **Open** button from the **Standard** group of the **Home** tab or choose **Menu > File > Open** from the **Top Border Bar**; the **Open** dialog box is displayed.

3.   Browse to **PCB_UPPER_CASE_top_###** in Tut 2 folder; **PCB_UPPER_CASE_top_###** is displayed in **File name** drop-down list, refer to Figure 4-22. Then, choose the **OK** button; the model is displayed, refer to Figure 4-23.

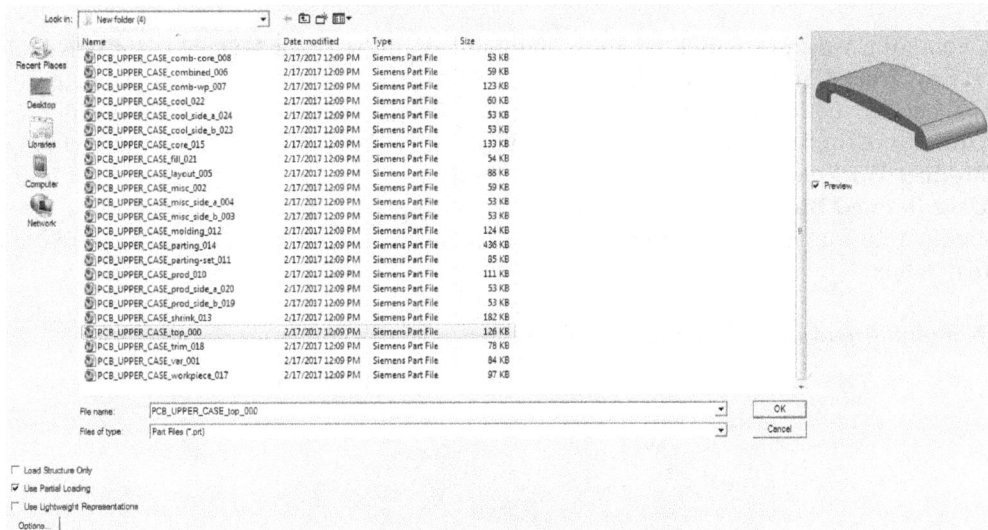

*Figure 4-22 The **Open** dialog box*

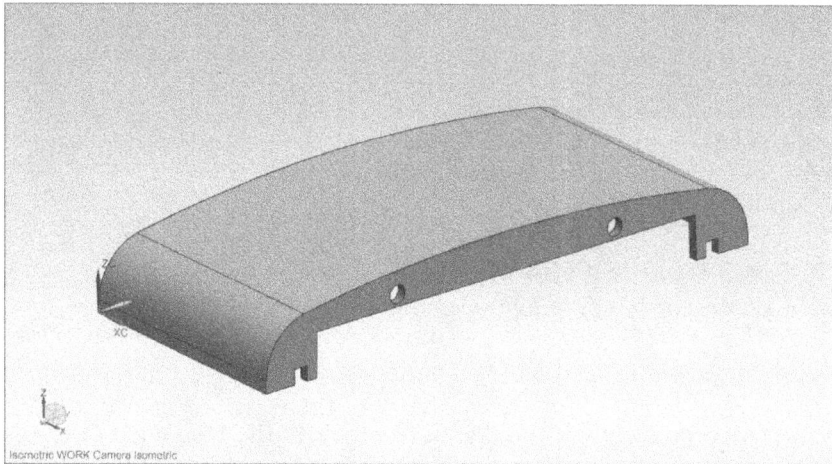

Figure 4-23 *The PCB_UPPER_CASE model*

Creating Workpiece

Now, you need to create the workpiece by using the **Workpiece** tool.

1. Choose the **Workpiece** tool from the **Main** gallery of the **Mold Wizard** tab; the **New Iray + Ray Traced Studio Rendering** window is displayed, refer to Figure 4-24. Choose the **OK** button; the **Workpiece** message box is displayed, refer to Figure 4-25(a). Choose the **OK** button; the **Workpiece** dialog box is displayed, refer to Figure 4-25(b). By default, the **Product Workpiece** option is selected in the drop-down list of the **Type** rollout and the **User Defined Block** option is selected in the **Workpiece Method** drop-down list. Refer to Figure 4-26 for the size of the core and cavity insert. Choose the **OK** button to close the dialog box.

    Note the size of insert is automatically defined.

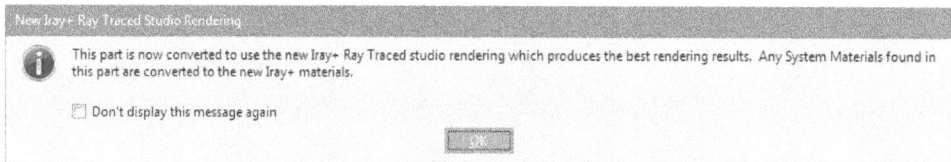

Figure 4-24 *The **New Iray + Ray Traced Studio Rendering** window*

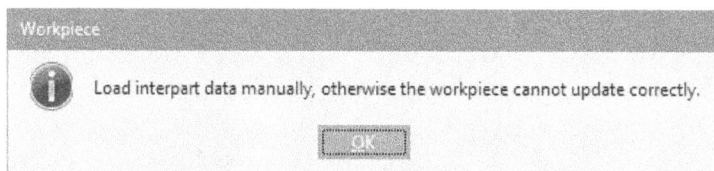

Figure 4-25(a) *The **Workpiece** message box*

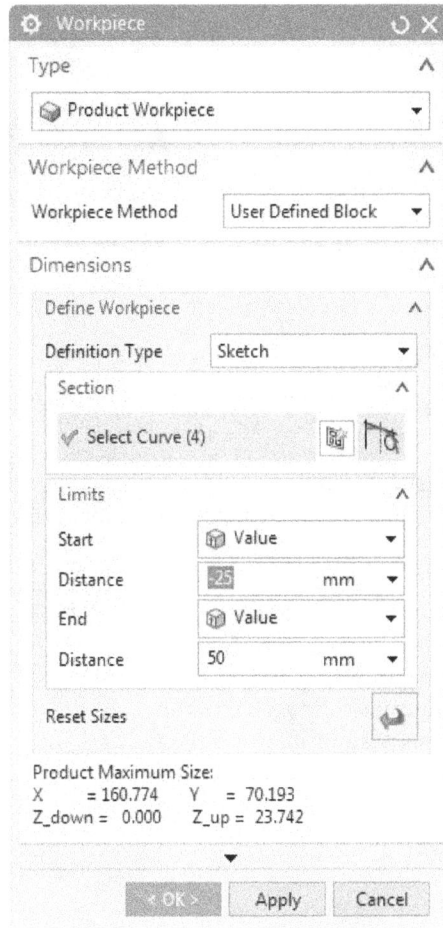

Figure 4-25(b) The **Workpiece** *dialog box*

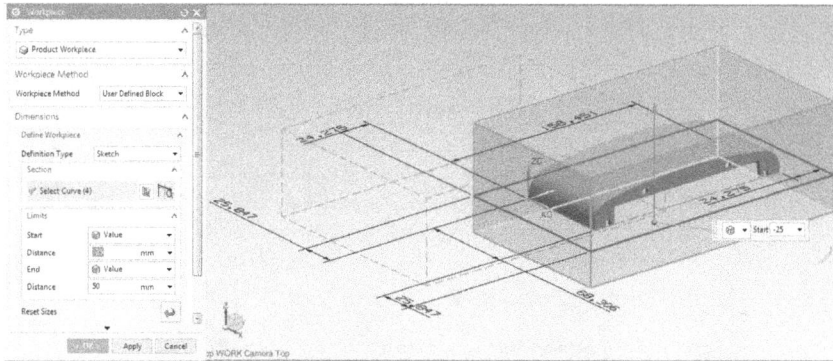

Figure 4-26 *The core and cavity insert*

Creating the Workpiece Layout

In this section, you will use the **Cavity Layout** tool to create the layout of the core and cavity insert.

1. Choose the **Cavity Layout** tool from the **Main** gallery of the **Mold Wizard** tab; the **Cavity Layout** dialog box is displayed, refer to Figure 4-27. The body of the core and cavity gets automatically selected and a green tick mark becomes visible in the **Select Body** area of the **Product** rollout. By default, the **Rectangular** option and the **Balanced** radio button are selected in the **Layout Type** rollout.

2. Select **2** from the drop-down list in the **Cavity Count** edit box of the **Balanced Layout Settings** rollout.

3. Click in the **Specify Vector** area of the **Layout Type** rollout and specify **XC** axis as the direction. Choose the **Start Layout** button from the **Generate Layout** rollout; the layout gets created, refer to Figure 4-28.

4. Choose the **Auto Center** button from the **Edit Layout** rollout to align the center of layout with the center of the mold base.

5. Choose the **Edit Insert Pocket** button from the **Edit Layout** rollout; the **Insert Pocket** dialog box is displayed, refer to Figure 4-29.

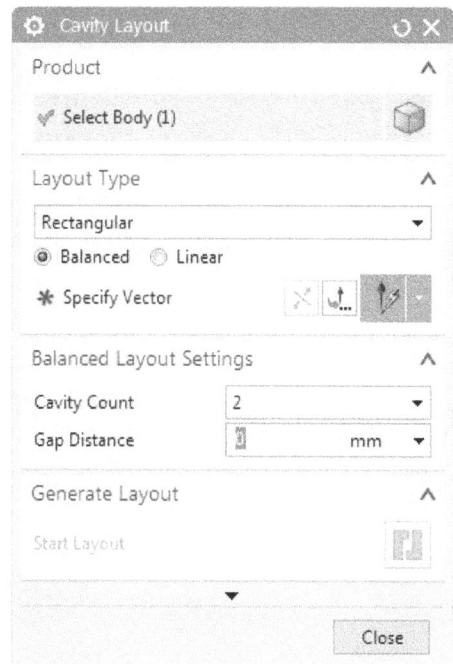

Figure 4-27 *The **Cavity Layout** dialog box*

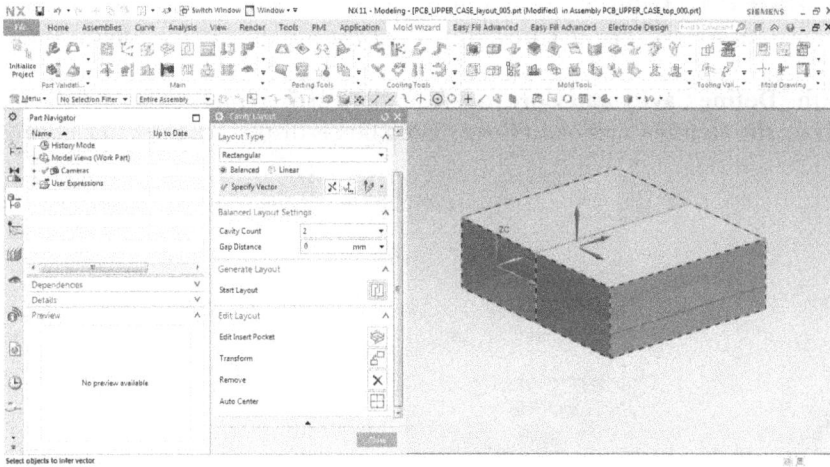

Figure 4-28 *The cavity layout of the workpiece*

Figure 4-29 *The **Insert Pocket** dialog box*

6. Select **5** from the **R** drop-down list and **2** from the **type** drop-down list; then choose the **OK** button to close the **Insert Pocket** dialog box. The **Cavity Layout** dialog box reappears.

7. Choose the **Close** button to close the **Cavity Layout** dialog box.

Creating Core and Cavity

In this section, you will create core and cavity by using the **Define Cavity and Core** tool.

1. Choose the **Define Cavity and Core** tool from the **Parting Tools** gallery of the **Mold Wizard** tab; the **Define Cavity and Core** dialog box is displayed, refer to Figure 4-30.

Figure 4-30 The **Define Cavity and Core** *dialog box*

By default, the **Region** option is selected in the drop-down list of the **Type** rollout and the **Cavity region** is highlighted in the **Select Sheets** rollout.

2. Choose the **Apply** button; the **New Iray + Ray Traced Studio Rendering** window is displayed, refer to Figure 4-31. Choose the **OK** button; the **View Parting Result** dialog box and the cavity of the model get displayed. Refer to Figure 4-32 for cavity of the component. Choose the **OK** button to close the dialog box; a green tick mark is displayed before the **Cavity region** in the **Select Sheets** rollout of the dialog box. Now, select the **Core region** area in the **Select Sheets** rollout of the dialog box and repeat the same procedure as for the cavity; the core is created, refer to Figure 4-33.

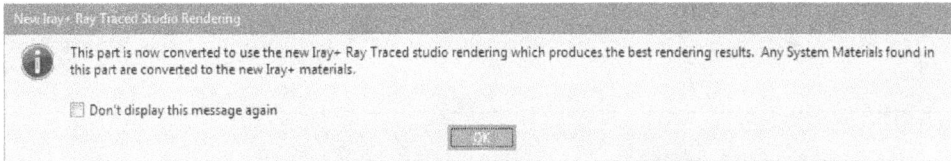

Figure 4-31 *The **New Iray + Ray Traced Studio Rendering** window*

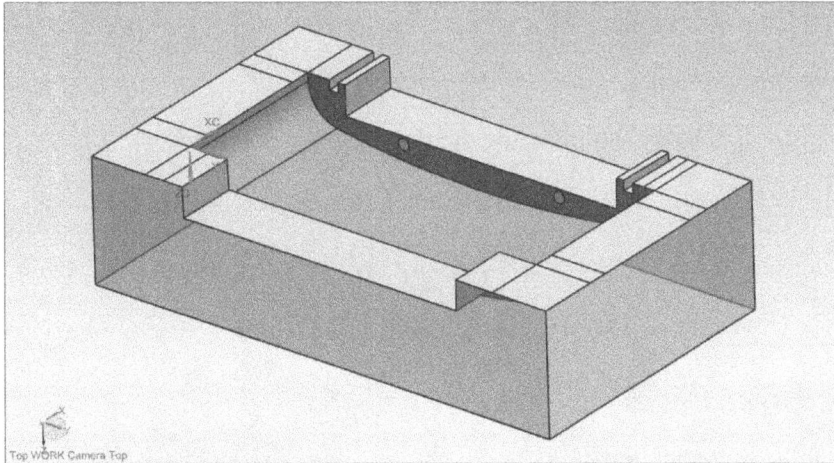

Figure 4-32 *The resultant cavity of the component*

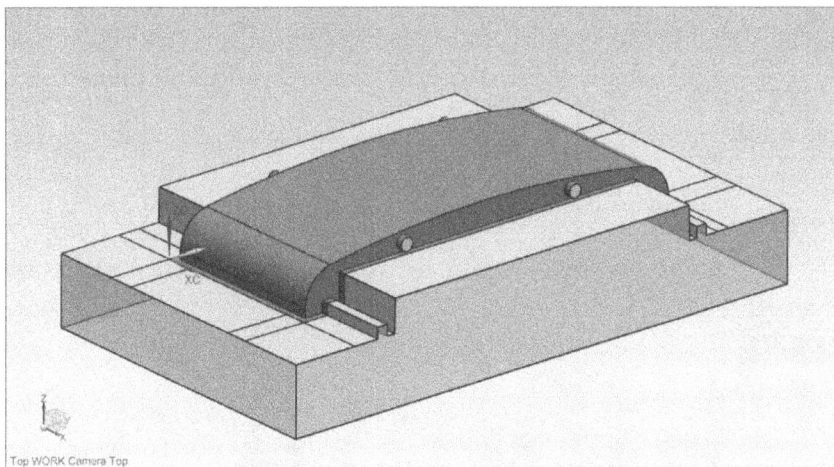

Figure 4-33 *The resultant core of the component*

3. Choose the **Cancel** button to close the dialog box.

4. Select the **PCB_UPPER_CASE_parting_###** in the assembly navigator and right-click on it; a shortcut menu is displayed. Select **PCB_UPPER_CASE_top_###** from the **Display Parent** shortcut menu.

Saving and Closing the File

1. Choose **Menu > File > Close > Save and Close** from the **Top Border Bar** to save and close the file.

Self-Evaluation Test

Answer the following questions and then compare them to those given at the end of this chapter:

1. Generally, core gives _____ profile to the component.

2. The _____ tool is used to define the the size of core and cavity.

3. The _____ option helps you to create the core and cavity only for single component.

4. The options in the _____ drop-down list are used to define the type of layout.

5. The _____ option is used to create a circular layout of the component.

Review Questions

Answer the following questions:

1. The _____ tool helps you to create the arrangement of layout of core and cavity workpiece.

2. The _____ option helps you to create the core and cavity for multiple component.

3. Generally, cavity gives _____ profile to the component.

4. The _____ tool helps you to create core and cavity of the component.

5. The _____ option are used to create rectangular layout of the workpiece.

EXERCISES

Exercise 1

In this exercise, you will open the output file (exr_01_top_###) of Exercise 1 of Chapter 3. Create the core and cavity for the model shown in Figure 4-34 and save it.   **(Expected time: 60 min)**

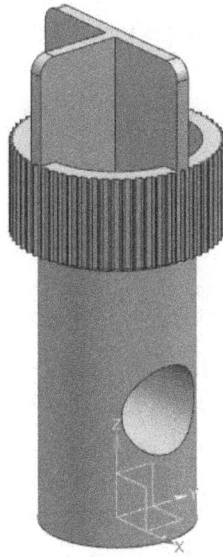

Figure 4-34 Model for Exercise 1

Exercise 2

In this exercise, you will open the outout file (exr_02_top_###) of Exercise 2 of chapter 3. Create the core and cavity for the model shown in Figure 4-35 and save it. **(Expected time: 60 min)**

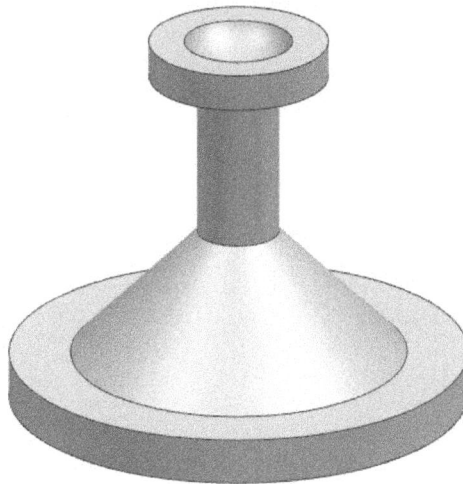

Figure 4-35 Model for Exercise 2

Answers to Self-Evaluation Test
1. inner, 2. Workpiece, 3. Product Workpiece, 4. Layout Type, 5. Circular

Chapter 5

Adding Mold Base and Standard Parts

Learning Objectives

After completing this chapter, you will be able to:

- *Understand plate structure*
- *Understand parts of plate structure and their selection*
- *Add mold base library*
- *Add locating ring*
- *Add sprue bush*

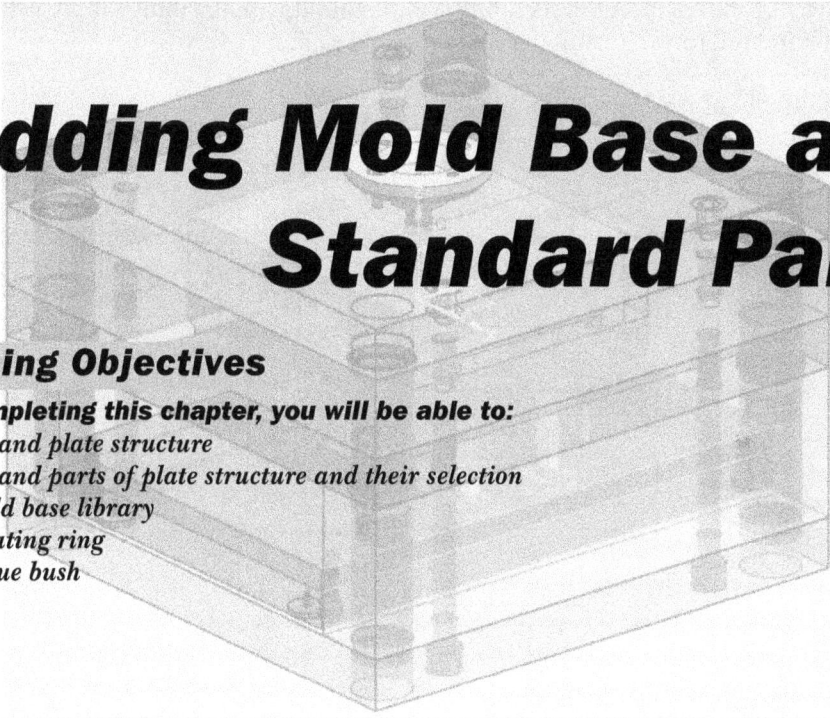

INTRODUCTION

In the previous chapter, you created core and cavity. After creating core and cavity, you need to add mold base and standard parts. In this chapter, you will learn how you can add mold base and standard parts using various tools.

MOLD BASE

Mold Base is a set of plates which holds the core and cavity of a component. Mold Base may be purchased either in kit form or in assembled form. Generally, the kit form does not contain the core and cavity of the component, therefore it is the mold manufacturer who decides whether to purchase it in kit form or to manufacture the mold plates.

There are two types of structures for mold base:
   (i) 2-plate mold
   (ii) 3-plate mold

Generally, 2-plate mold construction is the preferred design in industrial practice. Refer to Figures 5-1 and 5-2 for the structure of the 2-plate and 3-plate mold.

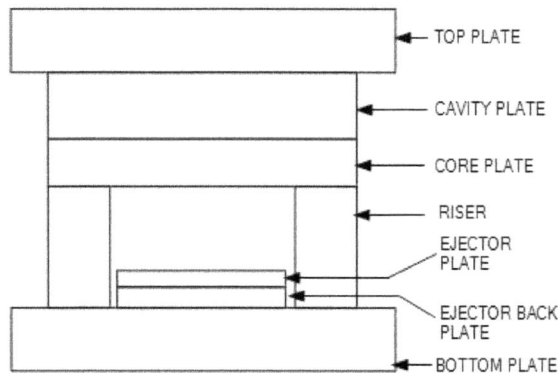

Figure 5-1 *The 2-plate structure of mold*

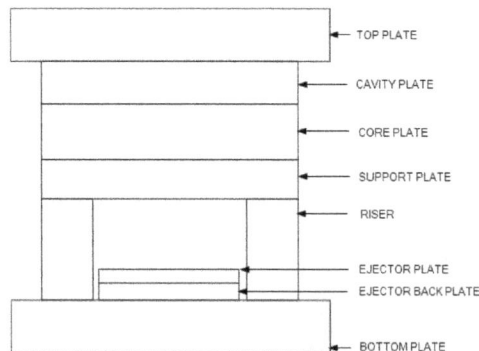

Figure 5-2 *The 3-plate structure of mold*

There are many suppliers in market who provide standard mold system to the customers. Brief details about the suppliers are as follows:

(i) **DME, Detroit Mold Engineering**: This company provides mold bases in various standard sizes and they also provide services on core and cavity machining. Refer to Figure 5-3 for DME mold system.

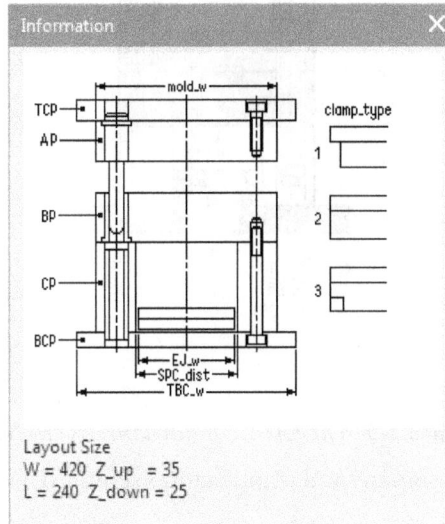

Figure 5-3 *The DME standard base in NX*

(ii) **DMS, Diemould Service Company Ltd**: This is a British company and it provides a range of mold bases and more than 12,000 products related to manufacturing industry. Refer to Figure 5-4 for DMS mold system.

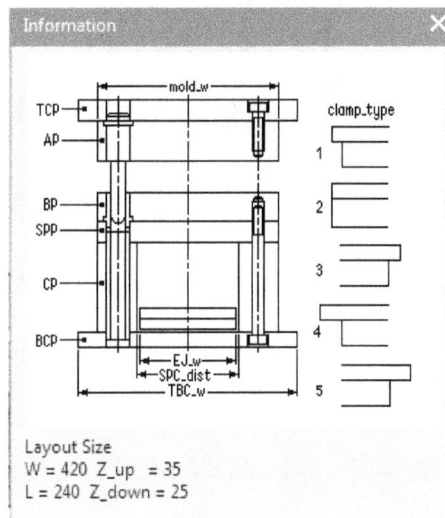

Figure 5-4 *The DMS standard base in NX*

(iii) **FUTABA**: It is a Japanese company and it originally produced vacuum tubes. Later, it entered into the mold manufacturing area. Refer to Figure 5-5 for Futaba mold system.

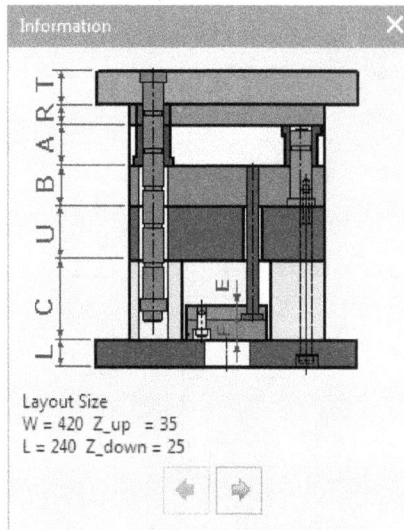

Figure 5-5 The FATUBA standard base in NX

(iv) **HASCO**: It is a German company and manufactures standard mold units. Refer to Figure 5-6 for HASCO mold system.

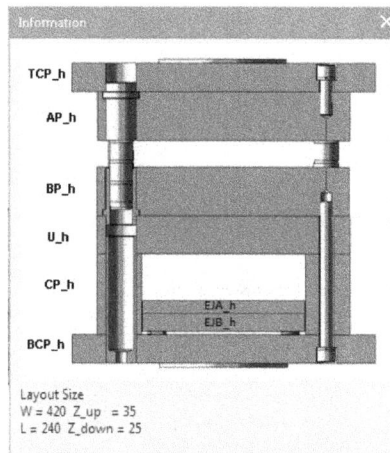

Figure 5-6 The HASCO standard base in NX

(v) **LKM**: It is a China based company. It is one of the suppliers of Sino mould. Refer to Figure 5-7 for LKM mold system.

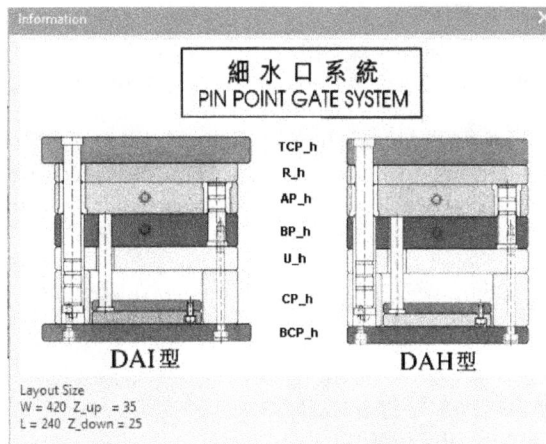

Figure 5-7 *The LKM standard base in NX*

(vi) **MEUSBURGER**: It is a Wolfurt, Austria based company. More than 17,000 customers all over the world make use of numerous advantages of standardization and benefits from the company. Refer to Figure 5-8 for MEUSBURGER mold system.

(vii) **RABOURDIN**: It is a leading french supplier of components for the mold and die industry. Refer to Figure 5-9 for RABOURDIN mold system.

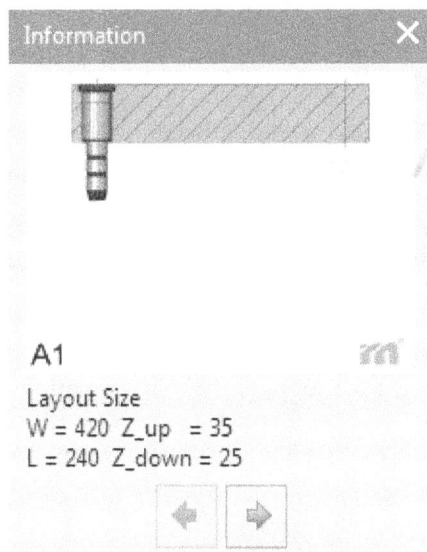

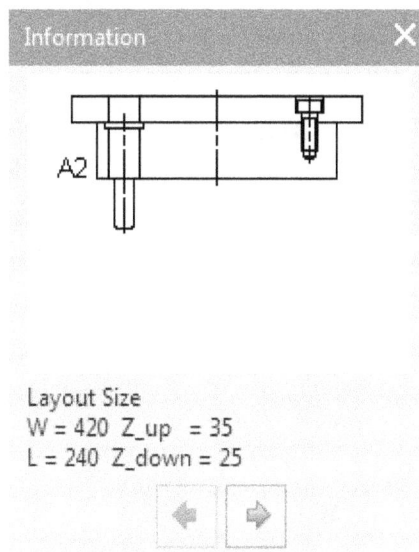

Figure 5-8 The MEUSBURGER standard base in NX

Figure 5-9 The RABOURDIN standard base in NX

Selection of Mold base

Generally, the selection of mold base depends on the selection of gate type. With a pin point gate, use 3-plate mold base and in case of side and tunnel gate, use 2-plate structure.

The material used for mold structure is carbon steel and it is used most often in non-hardened conditions. In special applications, prehardened steel, stainless steel, or aluminium alloy is also used. A mold base is used in combination with accessory parts like guide pillar, guide bush, and return pins.

The size selection of mold structure depends on the size of the component, insert, guide pillar, guide bush, sliders, ejection stroke, and so on.

Various Parts of Mold Structure

Generally, there are two halves in a mold:

(i) Fixed Half
(ii) Moving Half

Fixed Half mainly consists of Top Plate, Cavity Plate, and Guide Pillar.

Moving Half mainly consists of Bottom Plate, Core Plate, Riser, Ejector Plate, Ejector Back Plate, and Guide Bush.

Mold Base Library

Ribbon:	Mold Wizard > Main > Mold Base Library

The **Mold Base Library** tool helps you to add mold base assembly to the core and cavity. You can edit the mold base configuration and dimension. You can also edit existing template and register new template. Additionally, you can replace a mold base with different mold base and customize mold base models in library. Invoke this tool from the **Main** gallery of the **Mold Wizard** tab; the **Mold Base Library** dialog box is displayed alongwith the **Information** window, refer to Figures 5-10 and 5-11. Also, the **Reuse Library** navigator is displayed on the left side in the window. Select the mold base manufacturer from the **MW Mold Base Library** of **Reuse Library** navigator and then double-click on the mold base standard from the **Member Select** panel; the **Mold Base Library** dialog box along with the **Information** window will be displayed. After selecting the mold base from the **Member Select** panel, specify the size in the **Details** rollout of the **Mold Base Library** dialog box. Next, choose the **Apply** button to add the mold base to the assembly. Now, if the mold base needs to be rotated to 90 degrees, choose **Rotate Mold Base** button in the **Part** rollout of the dialog box. Choose the **OK** button to close the dialog box. Refer to Figure 5-12 for mold base arrangement.

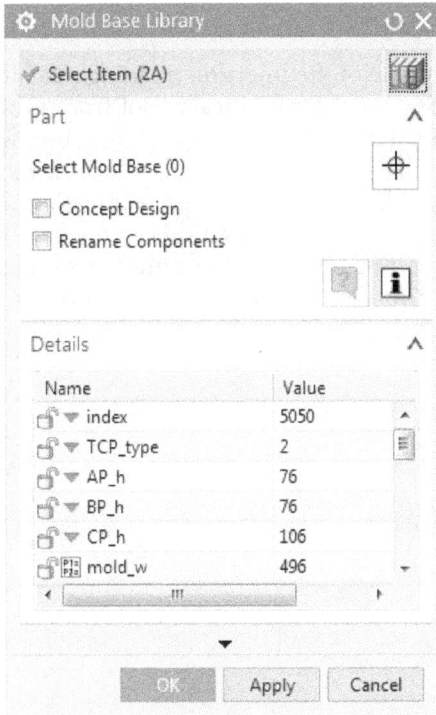

Figure 5-10 The **Mold Base Library**

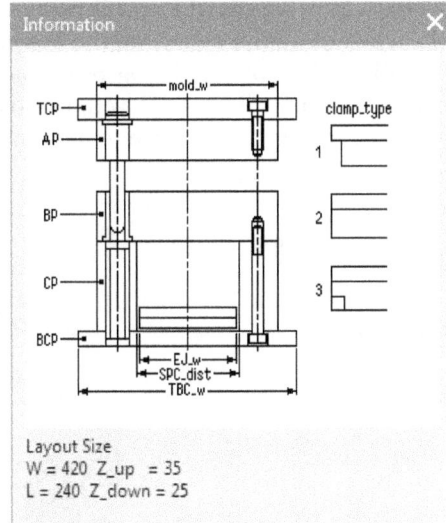

Figure 5-11 The **Information** window

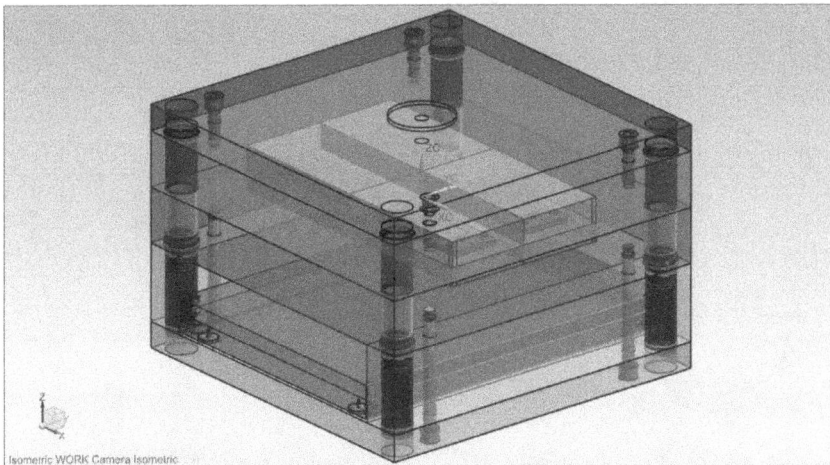

Figure 5-12 The Mold Base arrangement in NX Mold Wizard

Locating Ring

Ribbon: Mold Wizard > Main > Standard Part Library

Locating ring is used to locate the injection nozzle of the injection molding machine. To add locating ring to the mold, invoke the **Standard Part Library** tool from the **Main** gallery of the **Mold Wizard** tab; the **Standard Part Management** dialog box will be displayed alongwith the **Information** window, refer to Figures 5-13 and 5-14. Select the required standard from **MW Standard Part Library** in **Reuse Library** and then select **Injection folder**. Select the type of locating ring from the **Member Select** panel and then specify the parameters in the **Details** rollout of the dialog box. Choose the **OK** button to close the dialog box; the **Information** window will be displayed. Choose the **Close** button to close the window. Refer to Figure 5-15 for locating ring arrangement in mold assembly.

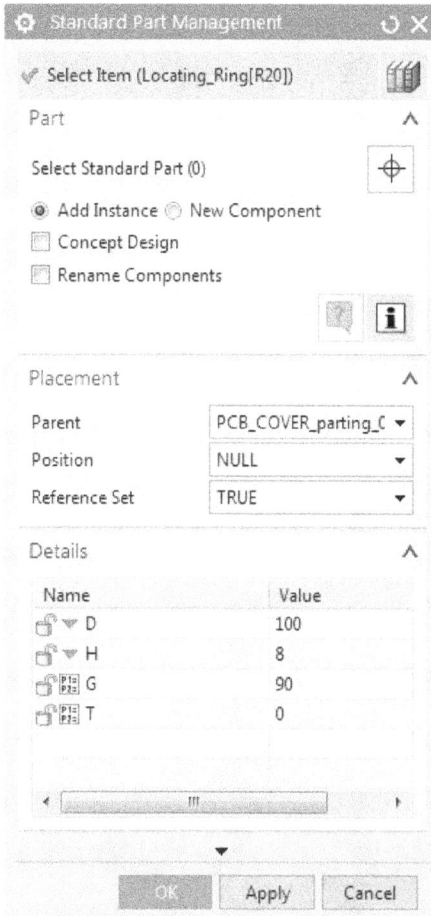

*Figure 5-13 The **Standard Part Management** dialog box*

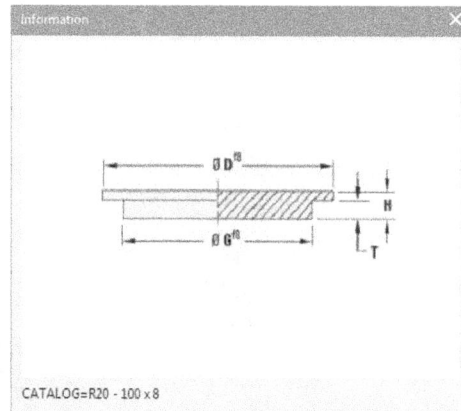

*Figure 5-14 The **Information** window*

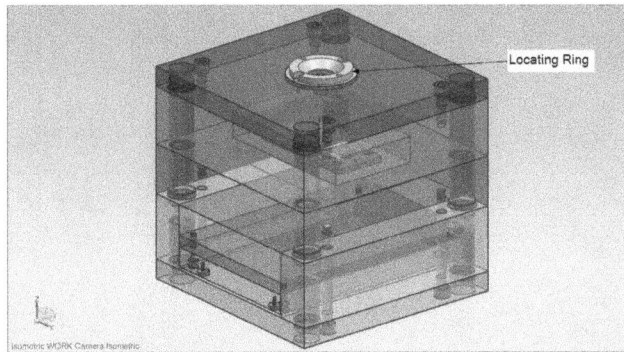

Figure 5-15 *The arrangement of locating ring in the mold assembly*

Sprue Bush

Ribbon: Mold Wizard > Main > Standard Part Library

Sprue Bush is a connecting member that connects the machine nozzle to the mold face, and provides suitable aperture through which the material can travel on its way to the mold cavity or the start of the runner in a multi-impression mold.

There are two basic designs for sprue bush depending on the form of seating between sprue bush and nozzle of machine:

    (i) Sprue Bush with spherical recess, refer to Figure 5-16.
    (ii) Sprue Bush has flat face, refer to Figure 5-17.

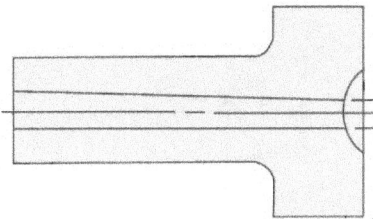

Figure 5-16 *The Sprue Bush with spherical recess*

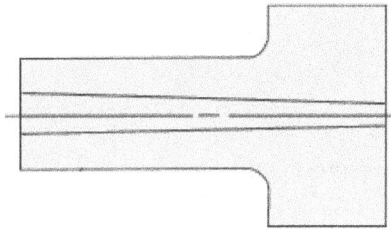

Figure 5-17 *The Sprue Bush with flat face*

Two main factors while selecting a sprue bush are:
(i) Depth and radius of sprue bush where nozzle contacts
(ii) Overall sprue bush length

To add a sprue bush to the mold base, invoke the **Standard Part Library** tool in the **Main** gallery of the **Mold Wizard** tab; the **Standard Part Management** dialog box will be displayed alongwith the **Information** window, refer to Figures 5-18 and 5-19. Select a standard from **MW Standard Part Library** in **Reuse Library** and then select the **Injection folder**. Select the type of the Sprue Bush from the **Member Select** panel and then specify the parameters in the **Details** rollout of the dialog box. Choose the **OK** button to close the dialog box. Refer to Figure 5-20 for Sprue Bush arrangement in mold assembly.

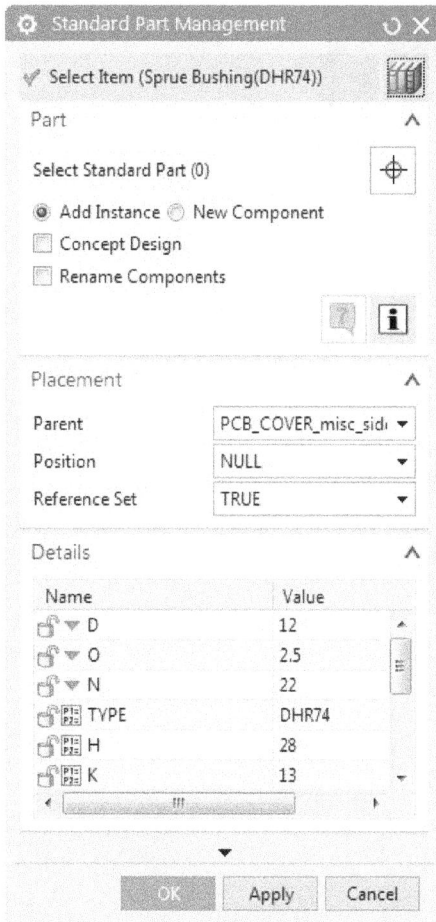

Figure 5-18 The **Standard Part Management** dialog box

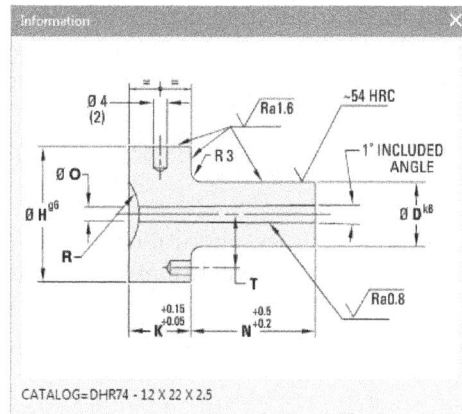

Figure 5-19 The **Information** window

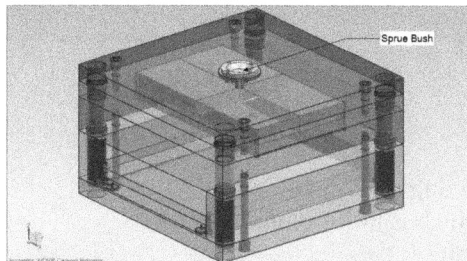

Figure 5-20 The arrangement of sprue bush in the mold assembly

TUTORIALS

To perform the tutorials, you need to download the zipped file named as *c05_NX_Mold_input* from the **Input Files** section of the CADCIM website. The complete path for downloading the file is:

> *Textbooks > CAD/CAM > NX_Mold > Mold Design using NX 11.0: A Tutorial Approach > Input Files*

After the file is downloaded, extract the folder. In this folder, you will find Tut1 and Tut2 folders containing input files for Tutorial 1 and Tutorial 2.

Tutorial 1

In this tutorial, you will add the mold base, register ring, and sprue bush to the model that was created in Tutorial 1 of Chapter 4.   **(Expected time: 2 hrs)**

The following steps are required to complete this tutorial:

a. Start NX and open the model, refer to Figure 5-21.
b. Add mold base to the model, refer to Figure 5-23.
c. Add register ring to the model, refer to Figure 5-25.
d. Add sprue bush to the model, refer to Figure 5-27.

Starting NX and Opening the Model

First, you need to start NX and then open a new file.

1. Double-click on NX shortcut icon available on the desktop of your computer to start NX.

2. Choose **Menu > File > Open** from the **Top Border Bar**; the **Open** dialog box is displayed.

3. Browse to **PCB_COVER_top_###** in Tut 1 folder; the **PCB_COVER_top_###** is displayed in the **File name** drop-down list. Next, choose the **OK** button; the model is displayed, refer to Figure 5-21.

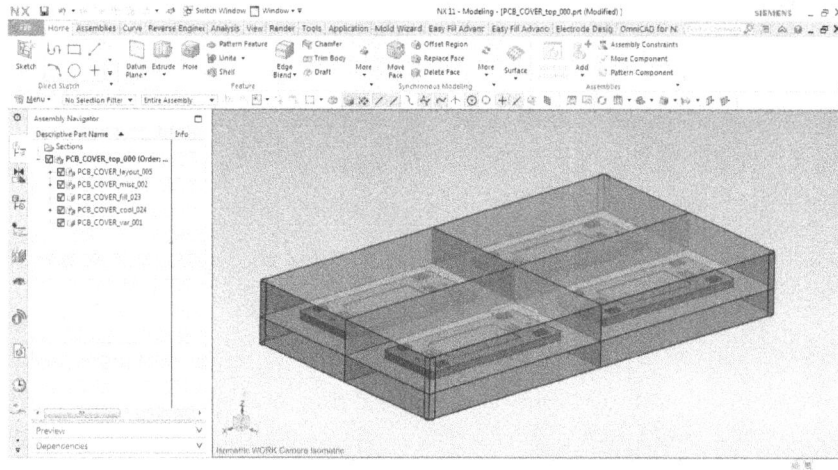

Figure 5-21 The PCB_COVER model for Tutorial 1

Adding Mold Base Assembly

Now, use the **Mold Base Library** tool to create plates for the mold assembly.

1. Choose **Reuse Library** from the **Resource Bar**.

2. Expand the **MW Mold Base Library** from the **Main** panel. Select the **DME** folder, right-click on **2A** in the **Member Select** panel and choose the **Insert** option from the shortcut menu; the **Mold Base Library** dialog box is displayed along with the **Information** window, refer to Figure 5-22 .

3. Select **5050**, **2**, **56**, **36**, and **106** from the **index**, **TCP_type**, **AP_h**, **BP_h**, and **CP_h** drop-down lists, respectively in the **Details** rollout. Next, choose the **OK** button to close the dialog box; the Mold plates are added to the assembly, refer to Figure 5-23.

4. Choose **File > Save > Save All** to save the file.

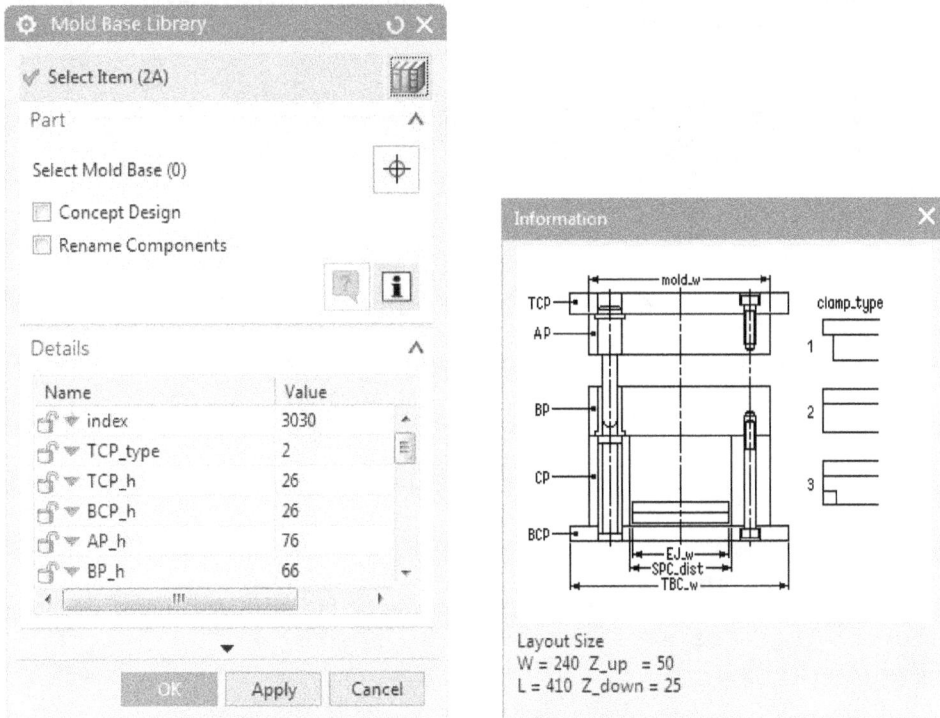

Figure 5-22 *The **Mold Base Library** dialog box and the **Information** window*

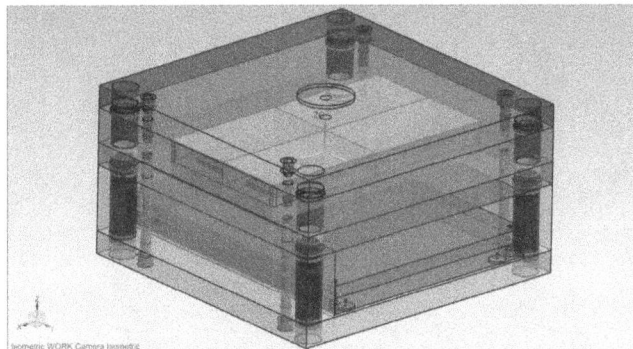

Figure 5-23 *The Mold Plate assembly*

Adding Register Ring to the Mold Assembly

Now, you need to add register ring to the mold assembly.

1. Choose **MW Standard Part Library** from **Reuse Library**. Expand **MW Standard Part Library** and then choose the type of locating ring from the **Injection** folder of the **DME_MM** folder.

2. Double-click on the **Locating_RING_With_Mounting_Holes[DHR21]** from the **Member Select** panel; the **Standard Part Management** dialog box is displayed with the **Information** window, refer to Figure 5-24.

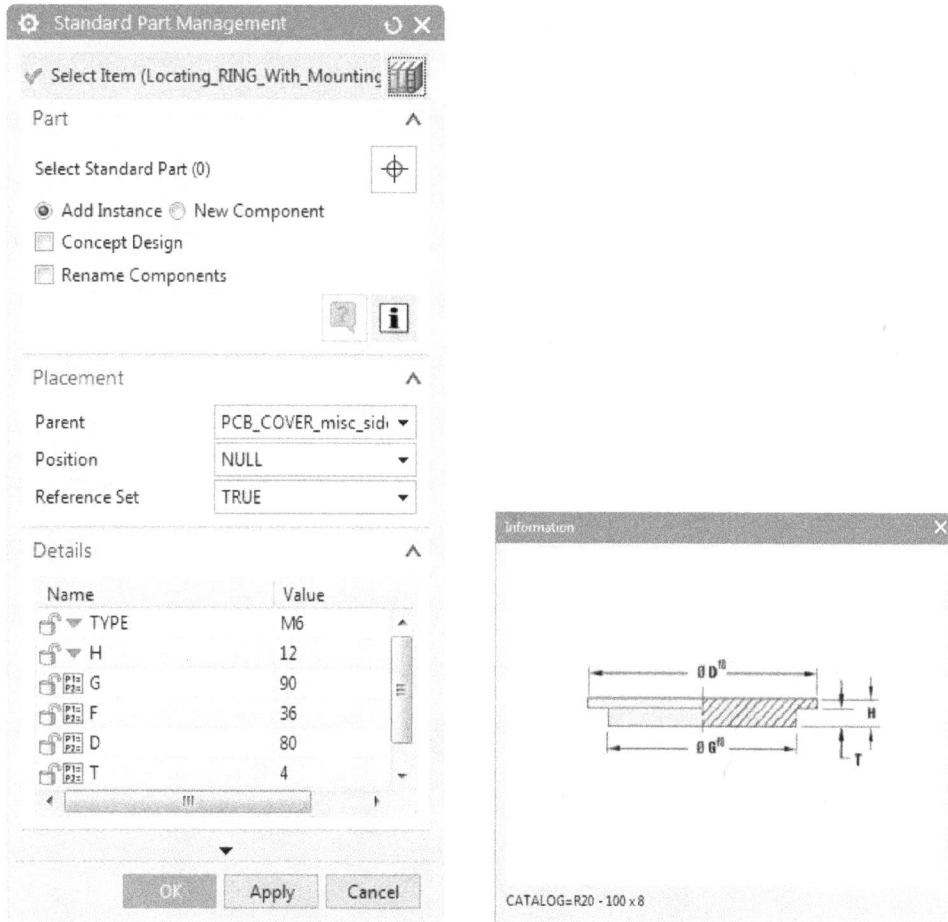

*Figure 5-24 The **Standard Part Management** dialog box and the **Information** window*

Note
Register ring and Locating ring refer to the same entity.

3. Select **M8** and **12** from the **Type** and **H** drop-down lists respectively. Enter **100** in the **D** edit box of the **Details** rollout and then choose the **OK** button; the **Information** window and the **Standard Part Management** dialog box is closed. Also, another **Information** window is displayed. Choose the **Close** button to close the window. Refer to Figure 5-25 for arrangement of register ring.

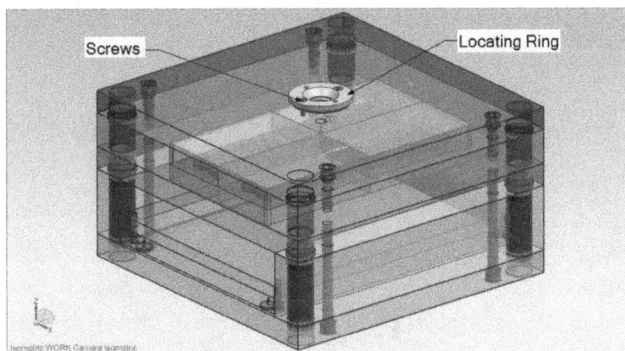

Figure 5-25 *The arrangement of Register Ring in the mold assembly*

Adding Sprue Bush to the Mold Assembly

Now, you need to add the sprue bush to the mold assembly.

1. Choose **MW Standard Part Library** from the **Reuse Library**. Expand the **MW Standard Part Library** and then choose type of sprue bush from the **Injection** folder in the **DME_MM** folder.

2. Double-click on the **Sprue Bushing(DHR74)** from the **Member Select** panel; the **Standard Part Management** dialog box is displayed along with the **Information** window, refer to Figure 5-26.

3. Select **18** and **4** from the **D** and **O** drop-down lists respectively. Enter **24** and **71** in the **K** and **N** edit boxes of the **Details** rollout.

   If the size of sprue bush is not appropriate then change it by double-clicking on the **Sprue Bushing (DHR74)** from the **Member Select** panel; the **Standard Part Management** dialog box is displayed. Click in the **Select Standard Part** area in the **Part** rollout and then select the sprue bush in the mold assembly. Now, you can change the size parameters or reposition the sprue bush. Refer to Figure 5-27 for arrangement of sprue bush in mold assembly.

4. Choose the **OK** button to close the dialog box.

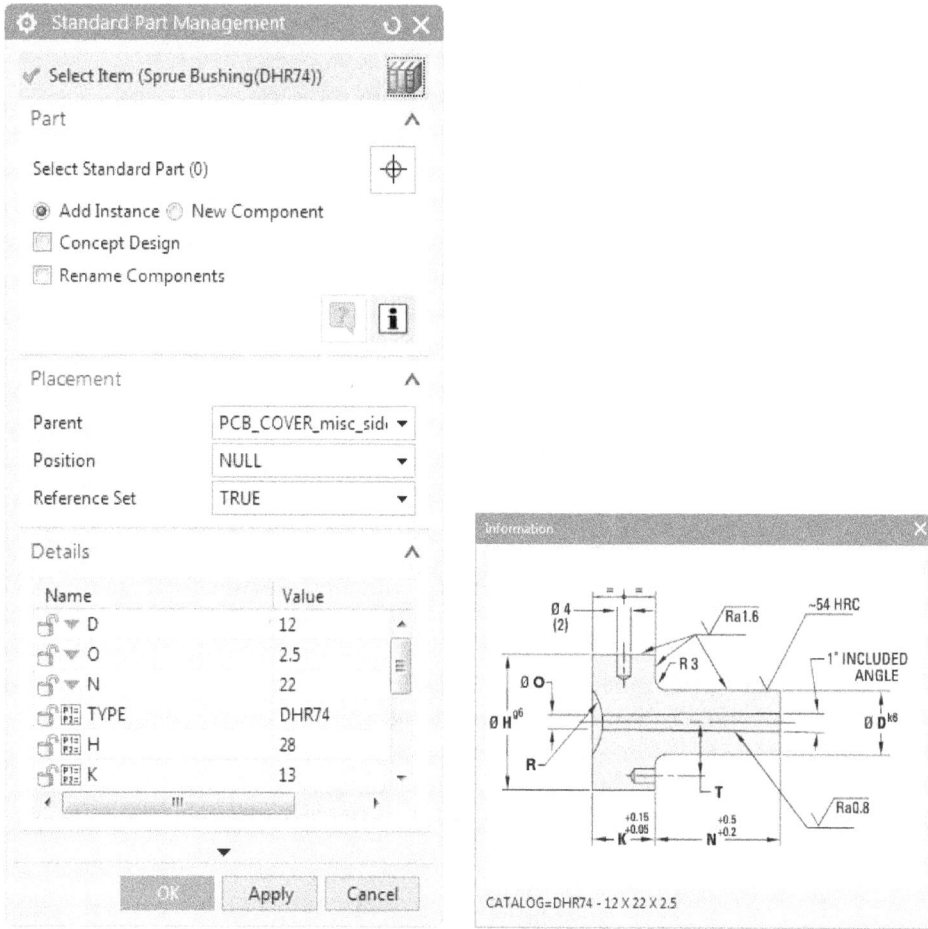

Figure 5-26 The **Standard Part Management** dialog box and the **Information** window

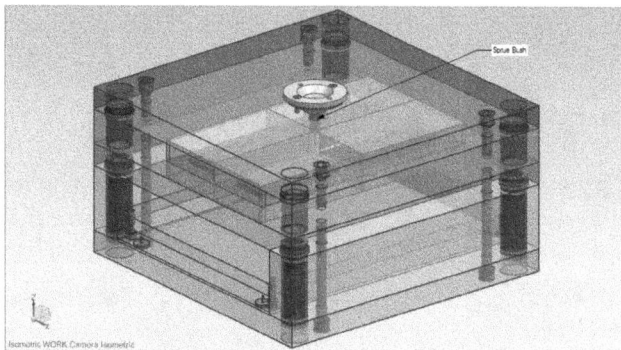

Figure 5-27 The arrangement of Sprue Bush in the mold assembly

Saving and Closing the File

1. Choose **Menu > File > Close > Save and Close** from the **Top Border Bar** to save and close the file.

Tutorial 2

In this tutorial, you will add the mold base, register ring, and sprue bush to the output file of Tutorial 2 of the previous chapter.									**(Expected time: 2 hrs)**

The following steps are required to complete this tutorial:

a. Start NX and open the model, refer to Figure 5-28.
b. Add mold base to the model, refer to Figure 5-30.
c. Add Register ring to the model, refer to Figure 5-32.
d. Add Sprue Bush to the model, refer to Figure 5-34.

Starting NX and Opening the Model

First, you need to start NX and then open a new file.

1. Double-click on the shortcut icon of NX available on the desktop of your computer.

2. Choose **Menu > File > Open** from the **Top Border Bar**; the **Open** dialog box is displayed.

3. Select the **PCB_UPPER_CASE_top_###** in Tut 2 folder; **PCB_UPPER_CASE_top_###** is displayed in the **File name** drop-down list. Now, choose the **OK** button; the model is displayed, refer to Figure 5-28.

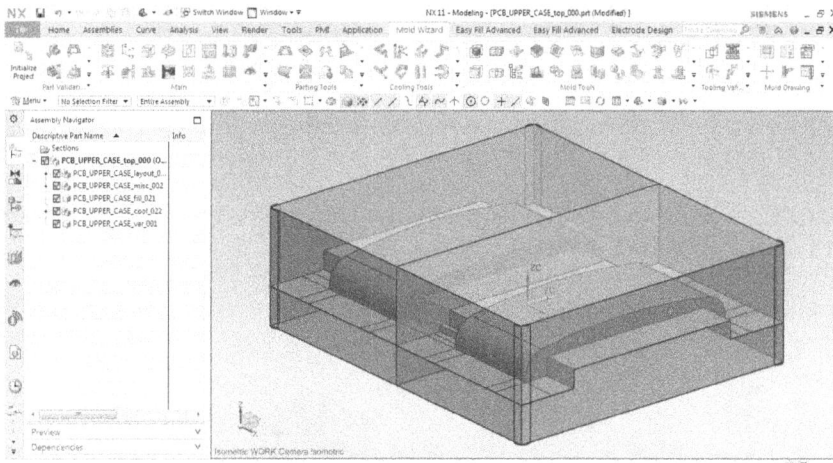

Figure 5-28 The PCB_UPPER_CASE_COVER model for Tutorial 2

Adding Mold Base Assembly

Now, use the **Mold Base Library** tool to create plates for the mold assembly.

1. Choose **Reuse Library** from the **Resource Bar**.

2. Expand **MW Mold Base Library** from the **Main** panel. Select the **DME** folder. Right-click on **2A** in the **Member Select** panel and choose the **Insert** option from the shortcut menu; the **Mold Base Library** dialog box is displayed along with the **Information** window, refer to Figure 5-29 .

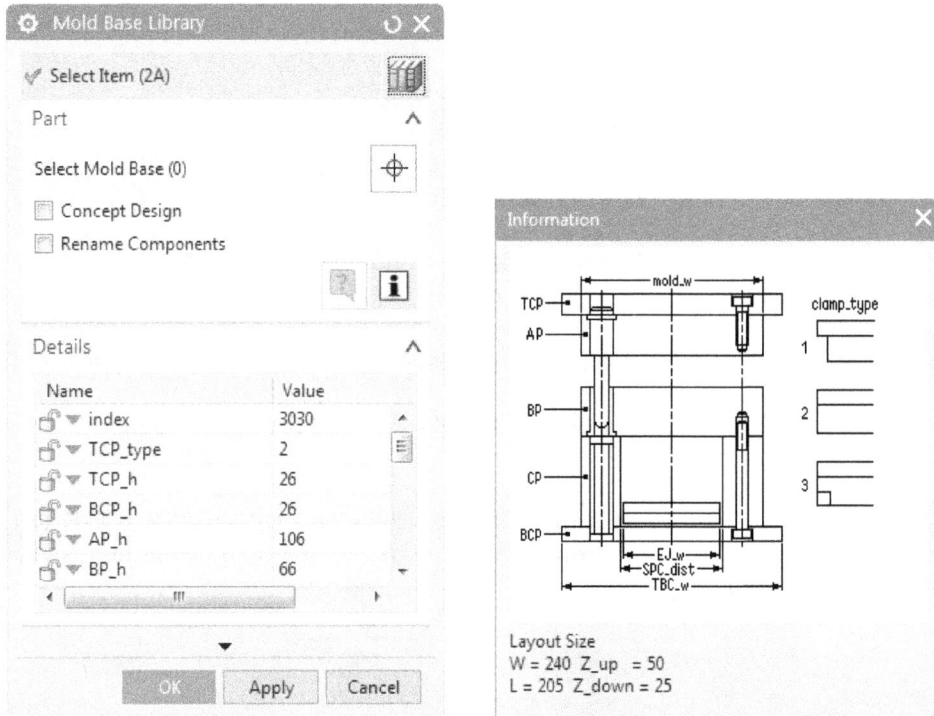

Figure 5-29 *The* **Mold Base Library** *dialog box and the* **Information** *window*

3. Select **5050**, **2**, **56**, **36**, and **106** from the **index**, **TCP_type**, **AP_h**, **BP_h**, and **CP_h** drop-down lists respectively in the **Details** rollout. Next, choose the **OK** button to close the dialog box; the mold plates are added to the assembly, refer to Figure 5-30.

4. Choose **File > Save > Save All** to save the file.

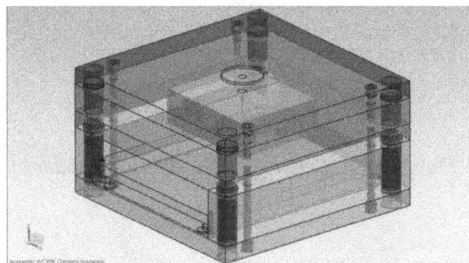

Figure 5-30 *The Mold Plate assembly*

Adding Register Ring to the Mold Assembly
Next, you need to add the register ring to the mold assembly.

1. Choose **MW Standard Part Library** from **Reuse Library**. Expand **MW Standard Part Library** and then choose the type of locating ring from the **Injection** folder in the **DME_MM** folder.

2. Double-click on **Locating_RING_With_Mounting_Holes[DHR21]** from the **Member Select** panel; the **Standard Part Management** dialog box is displayed along with the **Information** window, refer to Figure 5-31.

3. Select **M8** and **12** from the **Type** and **H** drop-down lists respectively. Ensure **100** is the value in the **D** edit box of the **Details** rollout and then choose the **OK** button; the **Information** window and the **Standard Part Management** dialog box are closed. Also, another **Information** window is displayed. Choose the **Close** button to close the window. Refer to Figure 5-32 for arrangement of register ring.

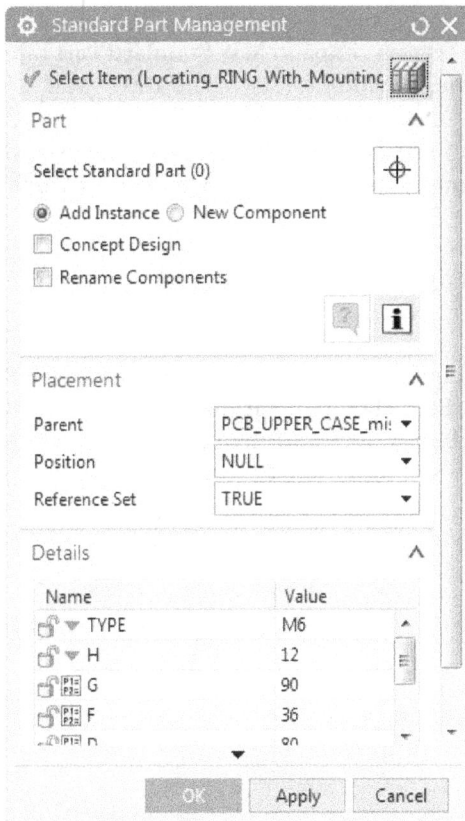

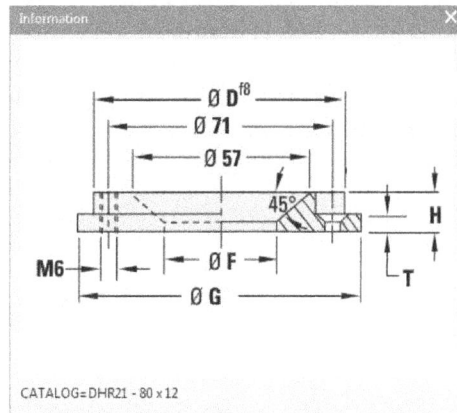

*Figure 5-31 The **Standard Part Management** dialog box and the **Information** window*

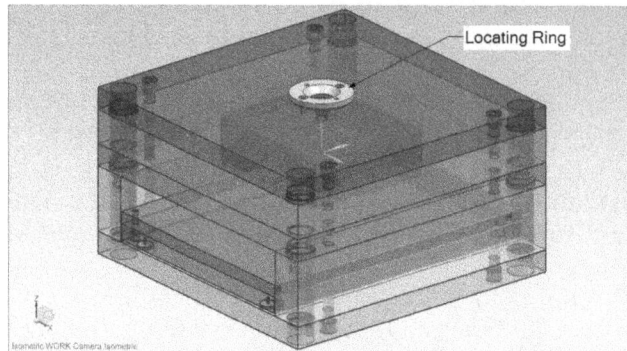

Figure 5-32 *The arrangement of register ring in the Mold assembly*

Adding Sprue Bush to the Mold Assembly

Now, you need to add the sprue bush to the mold assembly.

1. Choose **MW Standard Part Library** from **Reuse Library**. Expand **MW Standard Part Library** and then choose the type of sprue bush from the **Injection** folder in the **DME_MM** folder.

2. Double-click on **Sprue Bushing(DHR74)** from the **Member Select** panel; the **Standard Part Management** dialog box is displayed along with the **Information** window, refer to Figure 5-33.

3. Select **18** and **4** from the **D** and **O** drop-down lists respectively. Enter **24** and **71** in the **K** and **N** edit boxes of the **Details** rollout.

    If the size of sprue bush is not appropriate then double-click on the **Sprue Bushing (DHR74)** from the **Member Select** panel; the **Standard Part Management** dialog box is displayed. Click in the **Select Standard Part** area in the **Part** rollout and then select the sprue bush in the mold assembly. Now, you can change the size of parameters or reposition the sprue bush. Refer to Figure 5-34 for arrangement of sprue bush in the mold assembly.

4. Choose the **OK** button to close the dialog box.

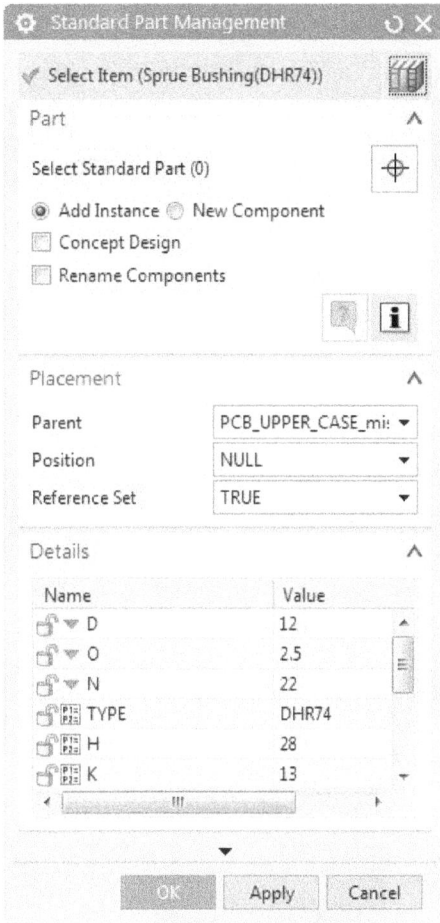

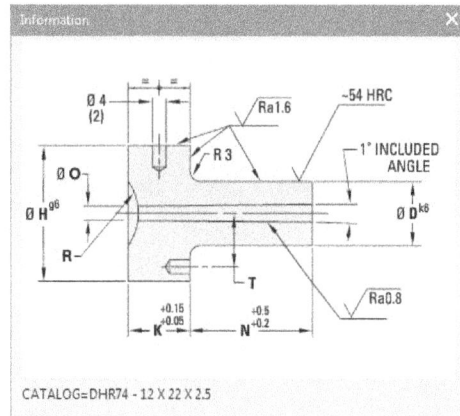

Figure 5-33 *The* ***Standard Part Management*** *dialog box and the* ***Information*** *window*

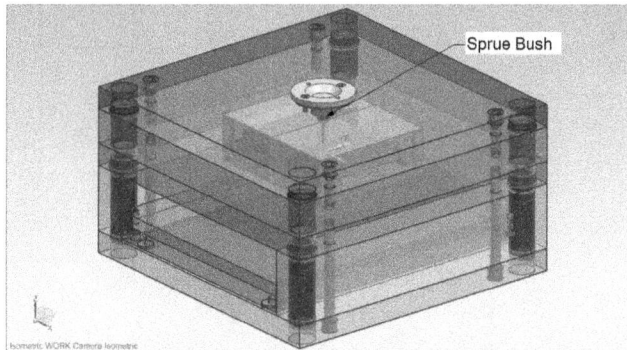

Figure 5-34 *The arrangement of Sprue Bush in the mold assembly*

Saving and Closing the File

1. Choose **Menu > File > Close > Save and Close** from the **Top Border Bar** to save and close the file.

Self-Evaluation Test

Answer the following questions and then compare them to those given at the end of this chapter:

1. Which of the following variables controls the height of the top plate?

    (a) **TCP_h**   (b) **BCP_h**
    (c) **EJA_h**   (d) None of these

2. Which of the following variables controls the height of the cavity plate?

    (a) **AP_h**   (b) **BP_h**
    (c) **CP_h**   (d) None of these

3. Which of the following variables controls the height of the core plate?

    (a) **AP_h**   (b) **BP_h**
    (c) **CP_h**   (d) None of these

4. Register ring is generally placed on the _____ plate.

5. You can orient the view of mold base using the _____ tool.

6. The _____ variable controls the diameter of register ring.

7. Sprue bush are of _____ types, _____ and _____.

8. The _____ tool is used to add mold base to a mold.

9. The _____ tool is used to add Sprue bush to a mold.

10. The _____ tool is used to add Register Ring to a mold.

Review Questions

Answer the following questions:

1. Which of the following dialog boxes is displayed when you choose the **Standard Part Library** tool from the **Main** gallery of the **Mold Wizard** tab?

    (a) **Standard**   (b) **Standard Part Management**
    (c) **Part Library**   (d) None of these

2. Which of the following fit types is used between guide pillar and guide bush?

    (a) **Press Fit**   (b) **Sliding Fit**
    (c) **Loose Fit**   (d) None of these

3. Register ring has _____ shape.

4. The _____ variable is used to control the guide pillar diameter.

5. The _____ variable is used to control the ejector plate width.

6. The _____ button is used to delete the standard parts from the mold assembly.

7. You can flip the direction of sprue bush. (T/F)

8. The **TCP_type** variable in the **Mold Base Library** dialog box is used to define the clamp type. (T/F)

9. You can replace a mold base with different one using the **Mold Base Library** tool. (T/F)

10. The Register ring helps to locate the mold in correct position on the injection machine. (T/F)

EXERCISES

Exercise 1

In this exercise, you will open the output file (exr_01_top_###) of Exercise 1 of Chapter 4. Add the mold base, register ring, and sprue bush to the model shown in Figure 5-35 and save it.

(Expected time: 1 hr)

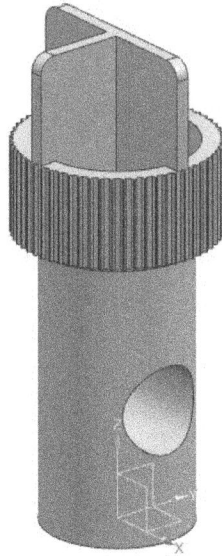

Figure 5-35 Model for Exercise 1

Exercise 2

In this exercise, you will open the output file (exr_02_top_###) of Exercise 2 of Chapter 4. Add the mold base, register ring, and sprue bush to the model shown in Figure 5-36 and save it.

(Expected time: 1 hrs)

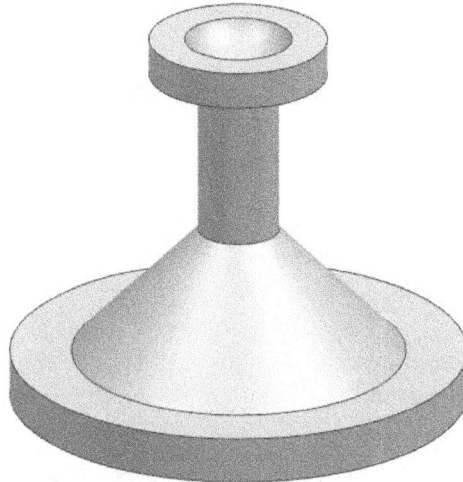

Figure 5-36 Model for Exercise 2

Answers to the Self-Evaluation Test
1. TCP_h, **2. AP_h**, **3. BP_h**, **4. Top**, **5. Mold CSYS**, **6.** D, **7.** two, Sprue Bush with spherical recess, Sprue Bush with flat face, **8. Mold Base Library**, **9. Standard Part Library**, **10. Standard Part Library**

Chapter 6

Creating Gate, Runner, and Layout

Learning Objectives

After completing this chapter, you will be able to:

• *Understand the feed system*
• *Define and position a gate*
• *Understand about different types of gates*
• *Create a gate*
• *Define runner and its layout*
• *Create a runner*

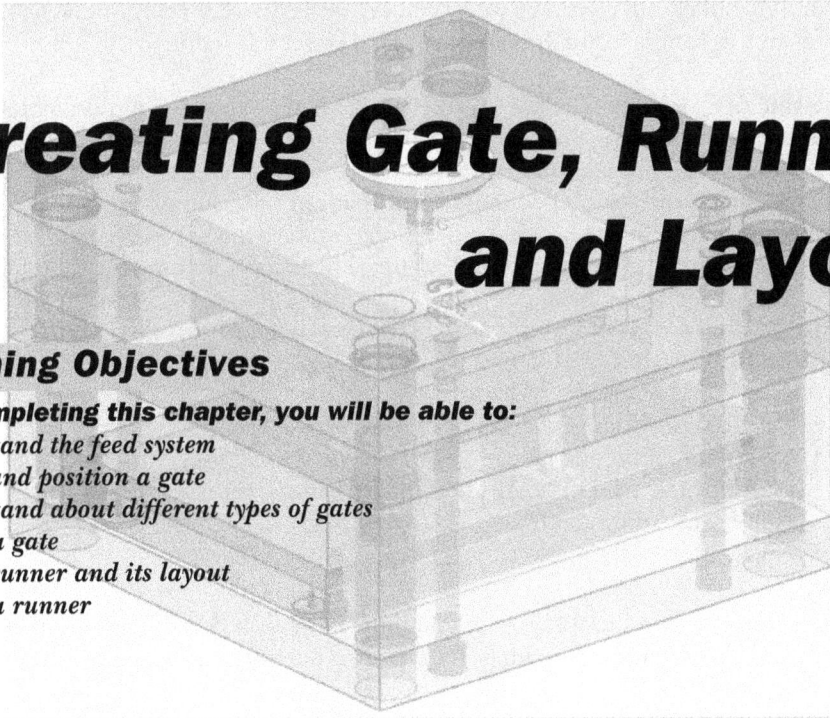

INTRODUCTION

In an injection mold, it is necessary to provide a flow-way between nozzle and impressions. This flow-way is also termed as the feed system. Normally, the feed system comprises a sprue, runner, and gate. You already learned to add sprue bush in the previous chapter. In this chapter, you will learn about the basic designing concepts of gate and runner, and then about the procedures to create the gate, runner, and runner layout.

GATE

Gate is a channel connecting a runner with the core or cavity of a component. It has smaller cross-sectional area as compared to the rest of the feed system. It is due to the following reasons:

(i)   The gate freezes soon after the impression is filled so that no void is created in the component during suck-back of injection plunger
(ii) It allows simple degating, and in some cases, it is automatic
(iii) After degating, only a small witness mark is left in the component
(iv) Better control of filling of multi-impressions can be achieved
(v)   Packing the core and cavity gap with material in excess for compensation of shrinkage

The size of a gate can be considered in terms of its cross-section and the length of gate. Gate length is also known as gate land. The optimum size of a gate depends on the following factors:

(i)   The flow characteristics of the material to be molded
(ii) The wall section of the mold
(iii) The volume of material to be injected into the impression
(iv) The temperature of the melt
(v)   The temperature of the mold

Ideally, the position of a gate should be such that there is an even flow of melt in the impression, fills uniformly, and also the advancing melt front spreads out and reaches the various extremities at the same time.

Types of Gate

The types of gate commonly used are:

(i)   Sprue gate
(ii)   Edge gate
(iii) Overlap gate
(iv)   Fan gate
(v)   Tab gate
(vi)   Diaphragm gate
(vii) Ring gate
(viii) Film gate
(ix)   Pin gate
(x)   Subsurface gate
(xi)   Winkle gate

Sprue Gate

When the mold is directly fed from a sprue or secondary sprue, the resultant feed section is known as a sprue gate, refer to Figure 6-1. The main disadvantage of this type of gate is that it leaves a large gate mark on the mold. The size of this mark depends on the following factors:

(i)   The diameter at the small end of the sprue
(ii) The sprue angle
(iii) The sprue length

To minimize the gate mark, you need to minimize the dimensions mentioned above.

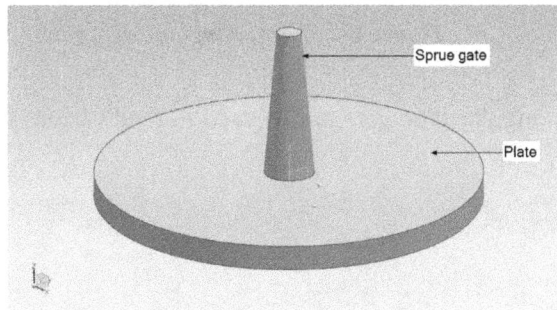

Figure 6-1 *The sprue gate*

Rectangular Edge Gate

It is a rectangular channel which is created by machining in the mold plate to connect the runner to the impression, refer to Figure 6-2. Advantages of this gate over other forms of gate are as follows:

(i)   The cross-sectional form is simple and therefore, easy to machine
(ii) Precise gate dimensions can be achieved
(iii) The gate dimensions can be easily and quickly modified
(iv) All common molding material can be molded through this type of gate

Disadvantage of this gate is that after gate removal, a witness mark is left on the visible surface of the resultant component.

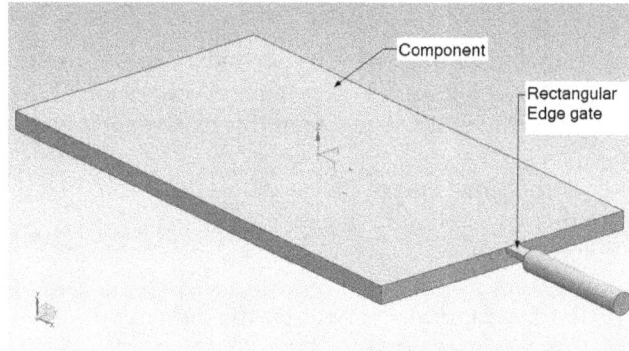

Figure 6-2 *The rectangular edge gate*

Overlap Gate

It is a variation of rectangular type gate and is used to feed various type of molding, refer to Figure 6-3.

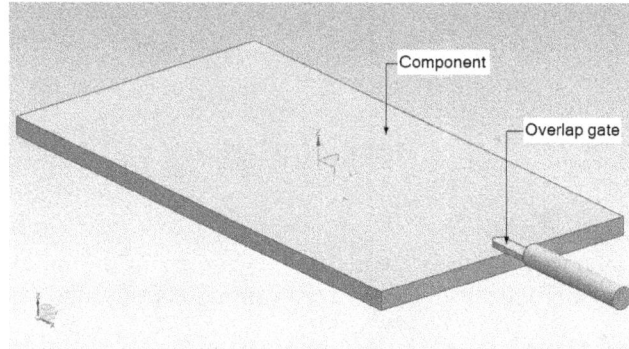

Figure 6-3 *The overlap gate*

Fan Gate

Fan gate is another edge-type gate but unlike the rectangular gate which has a constant width and depth, the corresponding dimensions of the fan gate are not constant. In this gate type, the width increases while the depth decreases so as to maintain a constant cross-sectional area throughout the length of the gate, refer to Figure 6-4.

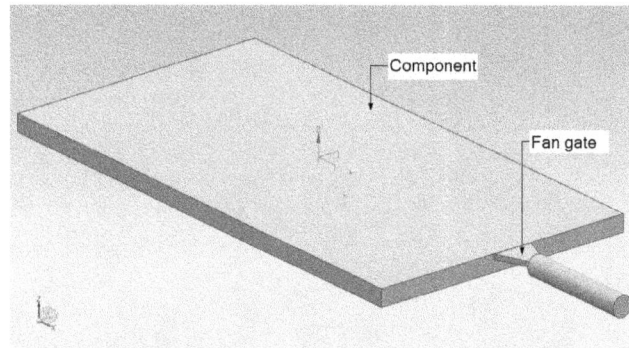

Figure 6-4 *The fan gate*

Tab Gate

This technique is used for creating solid block type molds. A projection or tab is molded on the side of the component, refer to Figure 6-5.

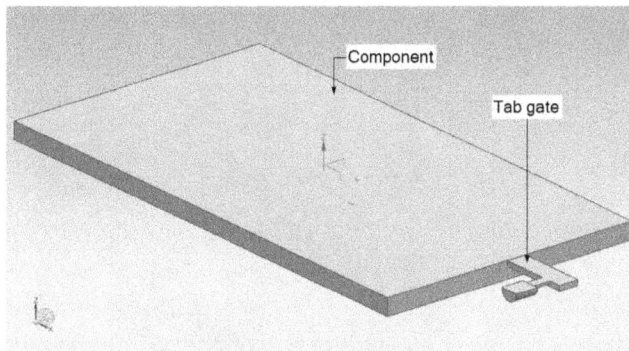

Figure 6-5 *The tab gate*

Diaphragm Gate

This gate is a long rectangular-type edge gate and it is used for large thin-walled components to assist the production of warpage free products, refer to Figure 6-6. This type of gate allows for constant filling of the mold and minimize the formation of weld lines.

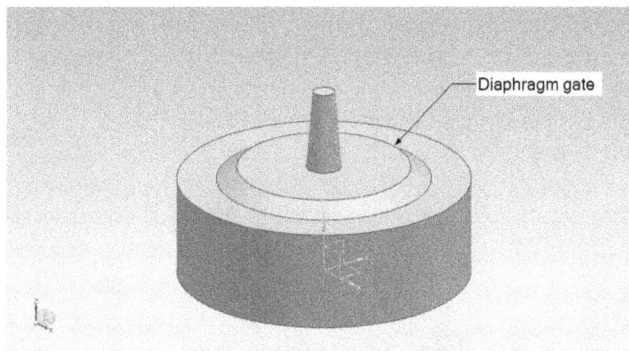

Figure 6-6 *The diaphragm gate*

Ring Gate

This type of gate is used for tubular-type moldings when more than one impression is required in a simple two-plate mold. This gate is used for all common molding materials. It is particularly useful for those materials which exhibit differential shrinkage and for which central feeding is not possible.

Pin Gate

This is a circular gate and is used for feeding into the base of components because it is relatively small in diameter, refer to Figure 6-7.

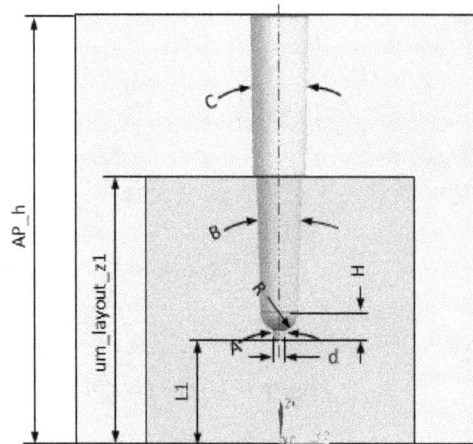

Figure 6-7 The pin gate

Round Edge Gate

This gate is formed by machining a matching semi-circular channel in both mold plates between the runner and the impression. Because of its form, the round edge gate suffers many disadvantages as compared to the rectangular edge gate, for example:

(i)   The matching form is more difficult to machine
(ii) Precise dimensions are more difficult to achieve

Because of these disadvantages, this gate is seldom used for creating molds having thickness below 4 mm.

Subsurface Gate

This gate is of circular or oval shape. It submerges and feeds into the impression below the parting surface of the mold, refer to Figure 6-8. Several advantages over the round gate are:

(i) No matching problems and precise dimensions can be achieved.
(ii) The gate is sheared from the molding during its ejection.

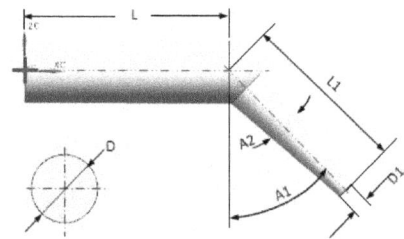

Figure 6-8 The subsurface gate

Winkle Gate

This is a curved variation of subsurface gate. This is also known as "curved subsurface gate" or "Curved tunnel gate", refer to Figure 6-9.

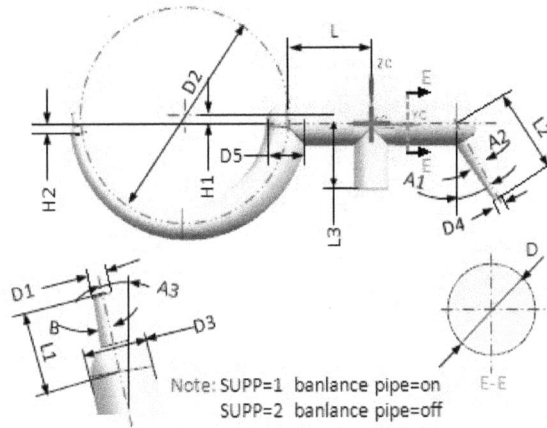

Figure 6-9 *The winkle gate*

Creating a Gate

Ribbon: Mold Wizard > Main > Design Fill

You can create a gate which connects with core or cavity. To create the gate, choose the **Design Fill** tool from the **Main** gallery of the **Mold Wizard** tab; the **Design Fill** dialog box will be displayed along with the **Information** window, refer to Figures 6-10 and 6-11. Also, the **Reuse Library** navigator will be displayed on the left in the window. Select the required type of gate from the **MW Fill Library** of the **Reuse Library** and then specify the parameters in the **Details** rollout of the **Design Fill** dialog box and then choose the **OK** button to close the dialog box.

The main parameters that controls the gate design is section type to be used. NX has various types of gate sections that help to control the amount of flow of material. The types of sections available are:

(a) Circular
(b) Parabolic
(c) Trapezoidal
(d) Hexagonal
(e) Semi_Circular

Depending upon the selection of the section type, the parameters to design the gate do vary. The main parameters are D (diameter) and L (length) which depend upon the selection of the section type.

If you want to edit the gate placement, choose the **Select Component** button from the **Select Component** area in the **Component** rollout of the **Design Fill** dialog box. Select the created gate from the model; a dynamic triad will be displayed. You can use this triad to move or rotate the gate. You can also modify the parameters of the gate in the **Details** rollout.

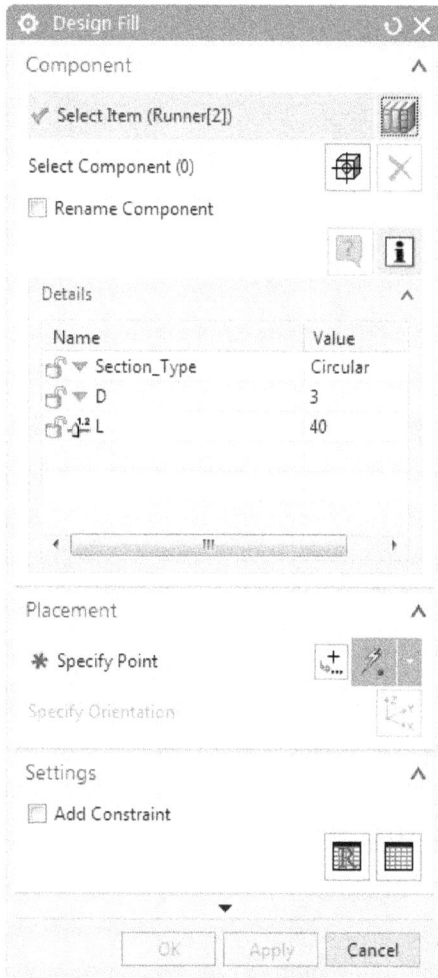

Figure 6-10 *The **Design Fill** dialog box*

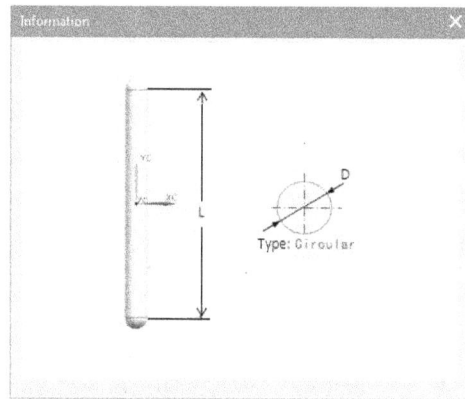

Figure 6-11 *The **Information** window*

RUNNER

Runner is a channel machined into the mold plates to connect the sprue with the gate to the impression. The wall of the runner channel must be smooth to prevent any restriction to flow. While designing a runner, a designer should consider the following points:

(i)   Shape of the cross-section of runner
(ii) Size of runner
(iii) The runner layout

Generally, in a mold four forms of cross-section of runner are used. They are as follows:

- (i)   Fully round/Circular
- (ii) Trapezoidal
- (iii) Modified Trapezoidal/Parabolic
- (iv) Hexagonal

The efficiency of a runner is equal to the ratio of cross-sectional area to perimeter. Following are the values of efficiency for some of the cross-sections:

Efficency of round runner = 0.25D
Efficiency of square runner = 0.25D
Efficiency of semicircular runner = 0.153D
Efficiency of rectangular runner = D/2 or D/4 or D/6

While designing the runner, a designer must consider the following factors:
- (i)   The wall section and the volume of molding
- (ii) The distance of impression from the main runner or sprue
- (iii) Cooling of runner
- (iv) Plastic material to be used
- (v)   Range of cutters available for moldmaker

Runner Layout
The layout of runner system depends upon the following factors:

- (i)   The number of impressions
- (ii) The shape of the components
- (iii) The type of mold
- (iv) The type of gate

While designing a runner layout, a designer should consider following points:

(i) The runner length should always be kept minimum to reduce pressure loss
(ii) The runner system should be balanced

It is not always possible to balance a runner system and this generally happens when you need to mold large number of impressions of different shapes. In such cases, you can achieve uniform filling of impression by varying the gate dimensions, which is referred as balanced gating. Figure 6-12 shows the examples of molds created based on balanced runner principle.

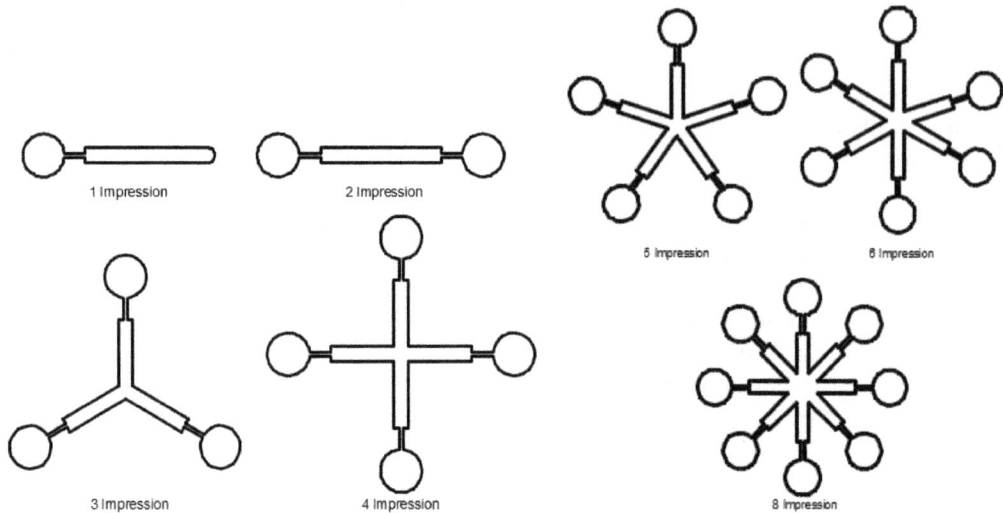

Figure 6-12 *balanced runner layout*

Creating a Runner (Method I)

Ribbon: Mold Wizard > Main > Design Fill

The **Design Fill** tool helps you to create a runner which connects with sprue bush and gate. To create a runner, choose the **Design Fill** tool from the **Main** gallery of the **Mold Wizard** tab; the **Design Fill** dialog box will be displayed, refer to Figure 6-13. Also, the **Information** window will be displayed, refer to Figure 6-14. Additionally, the **Reuse Library** navigator is displayed on the left in the window. Choose the **Point Dialog** button in the **Placement** rollout; you will be prompted to select the object. Select the position where you want to place the runner and then select the type of runner cross-section from the **Section Type** drop-down list in the **Details** sub-rollout of the **Component** rollout. Specify the parameters (D and L) of the runner and then choose the **OK** button to close the dialog box.

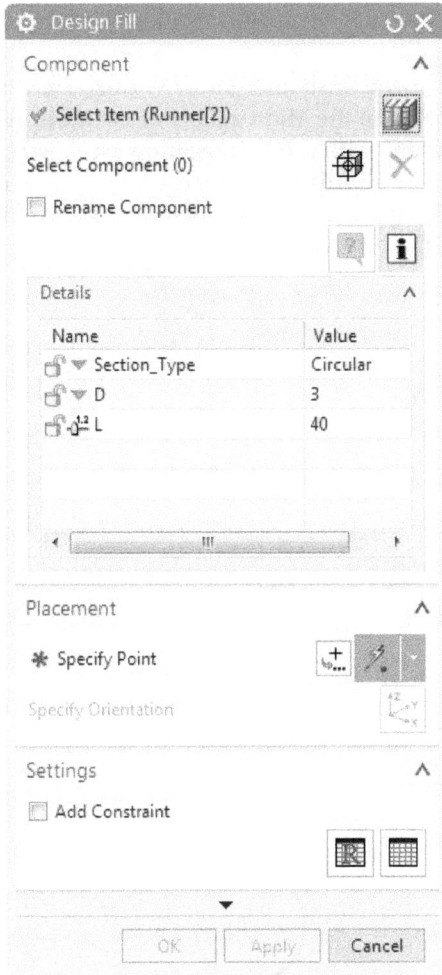

Figure 6-13 The **Design Fill** dialog box

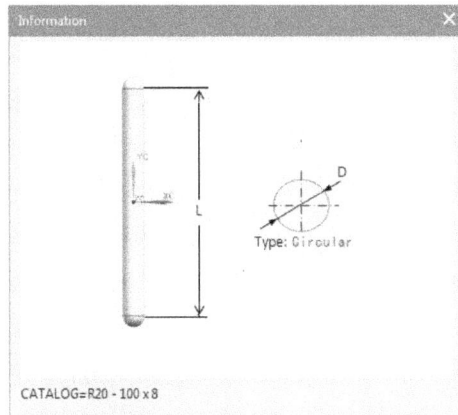

Figure 6-14 The **Information** window

The main parameter that controls the runner design is the section type to be used. NX has various types of runner sections available in it that help control the amount of flow of material. The section types available in NX are:

(a) Circular
(b) Parabolic
(c) Trapezoidal
(d) Hexagonal
(e) Semi_Circular

The parameters to design the runner vary depending upon the selection of the section type. The main parameters are D and L which depend upon the selection of the section type.

Creating a Runner (Method II)

Ribbon: Mold Wizard > Main > Runner

You can create a runner, which connects with sprue bush and gate, by using the **Runner** tool. To create a runner, choose the **Runner** tool from the **Main** gallery of the **Mold Wizard** tab; the **Runner** dialog box will be displayed along with the **Information** window, refer to Figures 6-15 and 6-16. Also, the **Reuse Library** navigator is displayed on the left in the window. Create the sketch of the runner layout and select the type of runner cross-section. Specify the parameters of the runner and then choose the **OK** button to close the dialog box.

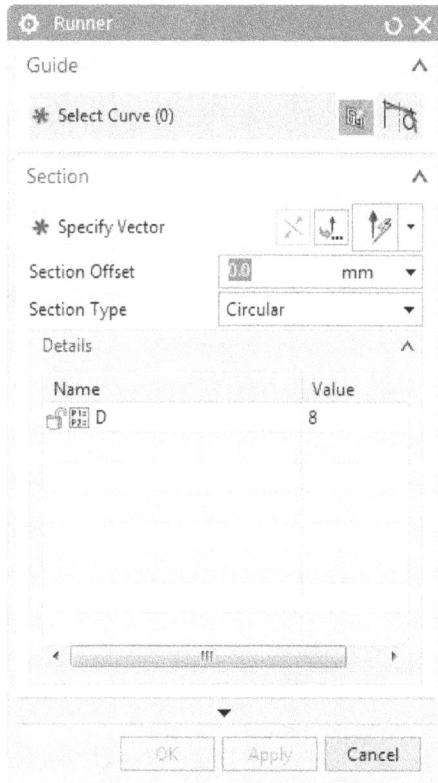

Figure 6-15 *The* **Runner** *dialog box*

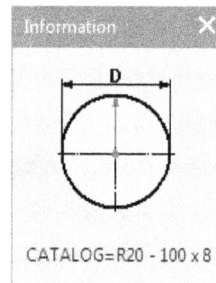

Figure 6-16 *The* **Information** *window*

TUTORIALS

To perform the tutorials, you need to download the zipped file named *c06_NX_Mold_input* from the **Input Files** section of the CADCIM website. The complete path for downloading the file is:

Textbooks > CAD/CAM > NX_Mold > Mold Design using NX 11.0: A Tutorial Approach > Input Files

After the file is downloaded, extract the folder. In this folder, you will find Tut1 and Tut2 folders containing input files for Tutorial 1 and Tutorial 2.

Tutorial 1

In this tutorial, you will create a gate and a runner to the model (PCB_COVER_top_###) contained in Tut 1 folder that you have downloaded. After creating the gate and runner, save the file.

(Expected time: 1 hr)

The following steps are required to complete this tutorial:

a. Start NX and open the model, refer to Figure 6-17.
b. Create gate, refer to Figure 6-19.
c. Create runner, refer to Figure 16-21.

Starting NX and Opening the Model

First, you need to start NX and then open a new file.

1. Double-click on NX shortcut icon on the desktop of your computer to start NX.

2. Choose the **Open** button from the **Standard** group of the **Home** tab or choose **Menu > File > Open** from the **Top Border Bar**; the **Open** dialog box is displayed.

3. Browse to **PCB_COVER_top_###** in Tut 1 folder; **PCB_COVER_top_###** is displayed in the **File name** drop-down list. Next, choose the **OK** button; the model is displayed, refer to Figure 6-17.

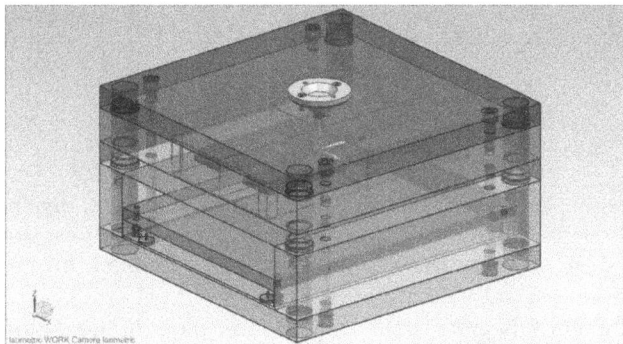

Figure 6-17 *The mold assembly*

Creating the Gate

Now, you will create the gate for the mold.

1. Choose the **Design Fill** tool from the **Main** gallery of the **Mold Wizard** tab; the **Design Fill** dialog box is displayed with the **Information** window, refer to Figure 6-18.

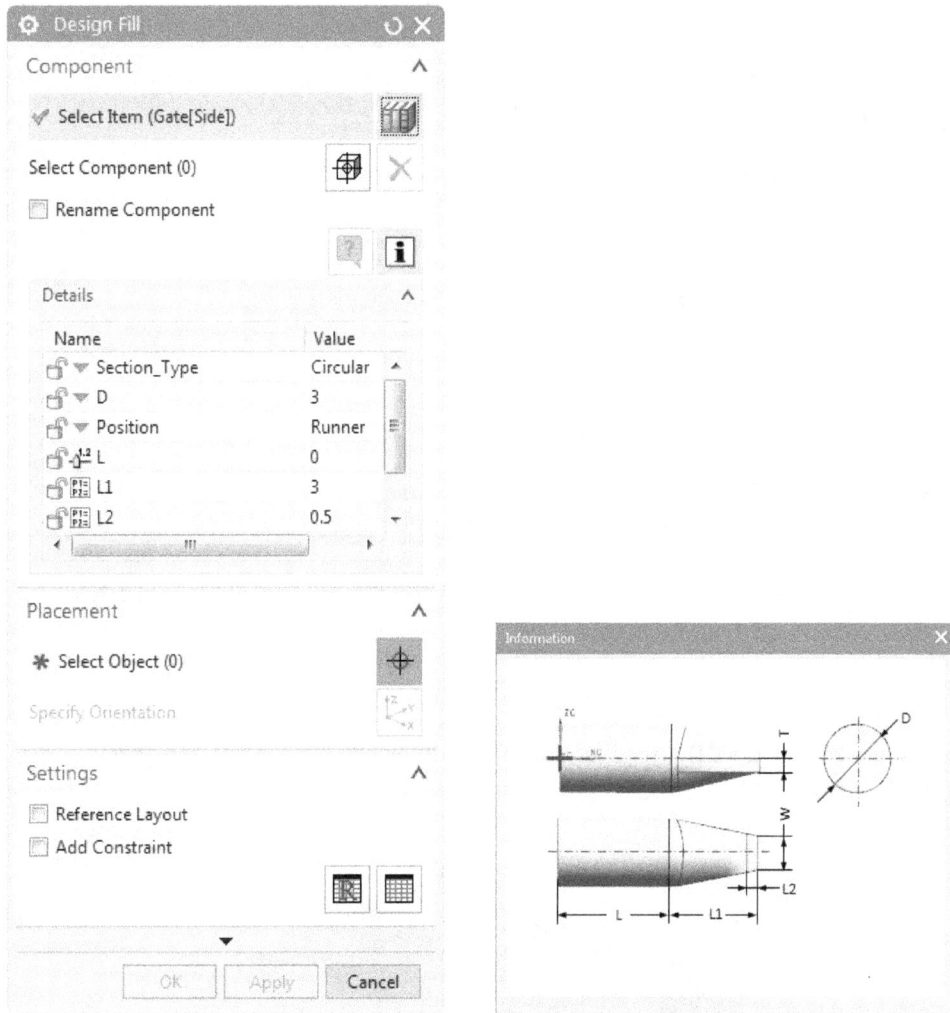

*Figure 6-18 The **Design Fill** dialog box and the **Information** window*

2. Select **Gate[Fan]** from the **Member Select** panel in the **Reuse Library**.

3. Specify the parameter of the gate in the **Details** sub-rollout of the dialog box. Make sure the value for **Section_Type** is **Circular** and **D** is **3**. Select the **Select Object** area in the **Placement** rollout and specify the gate position and orientation of gate, refer to Figures 6-19 and 6-20. Repeat the steps for placing other gates. After placing all the gates, choose the **OK** button to close the dialog box.

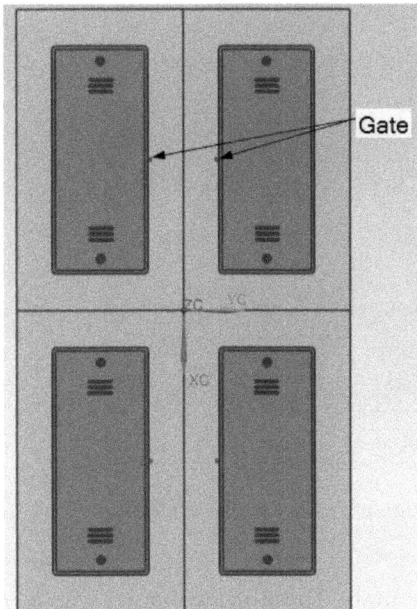

Figure 6-19 The gate position and orientation in the model

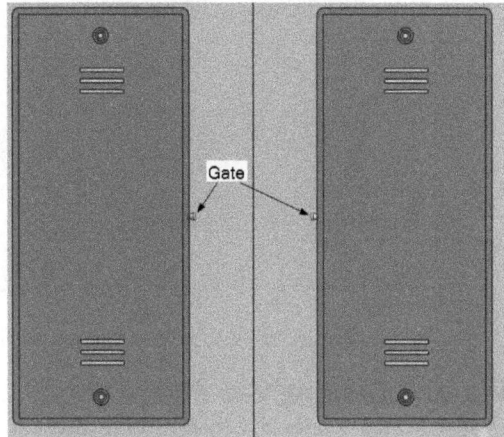

Figure 6-20 A zoom view of gate position in the model

Tip
For better visibility, you can hide solid bodies.

Creating the Runner

In this section, you will create a runner for mold component.

1. Choose the **Design Fill** tool from the **Main** gallery of the **Mold Wizard** tab; the **Design Fill** dialog box is displayed.

2. Select **Runner[4]** from the **Member Select** panel in the **Reuse Library**.

3. Select the **Circular** and **4.5** from the **Section_Type** and **D1** drop-down lists. Enter **205** and **45** in the **L1** and **L** edit boxes, respectively, in the **Details** sub-rollout of the dialog box. Select the **Specify Point** area in the **Placement** rollout and position the runner, refer to Figure 6-21.

4. After placing the runner, choose the **OK** button to close the dialog box.

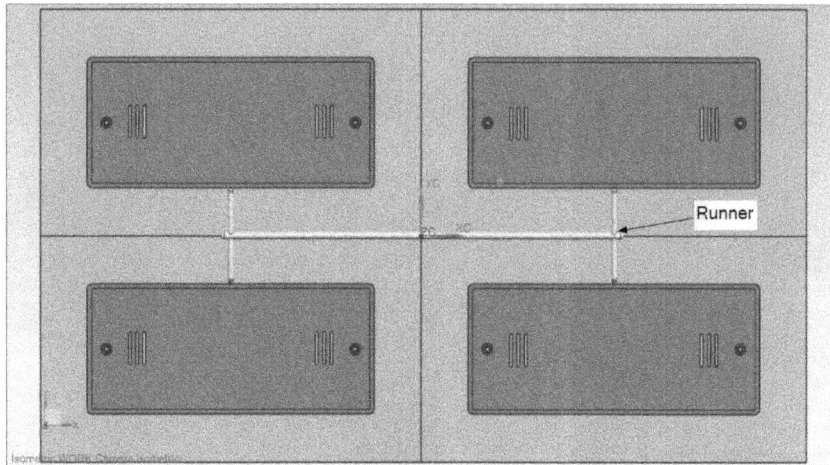

Figure 6-21 *The position and orientation of runner in the model*

Saving and Closing the File

1. Choose **Menu > File > Close > Save and Close** from the **Top Border Bar** to save and close the file.

Tutorial 2

In this tutorial, you will create a gate and a runner to the model (PCB_UPPER_CASE_top_###) contained in Tut 2 folder that you have downloaded. After creating the gate and runner, save the file.   **(Expected time: 1 hr)**

The following steps are required to complete this tutorial:

a. Start NX and open the model, refer to Figure 6-22.
b. Create the gate, refer to Figure 6-24.
c. Create the runner, refer to Figure 6-25.

Starting NX and Opening the Model

In this section, you will start NX and then open a new file.

1. Double-click on NX shortcut icon on the desktop of your computer to start NX.

2. Choose the **Open** button from the **Standard** group of the **Home** tab or choose **Menu >
    File > Open** from the **Top Border Bar**; the **Open** dialog box is displayed.

3. Browse to **PCB_UPPER_CASE_top_###** in Tut 2 folder; **PCB_UPPER_CASE_top_###**
    is displayed in the **File name** drop-down list. Then choose the **OK** button; the model is
    displayed, refer to Figure 6-22.

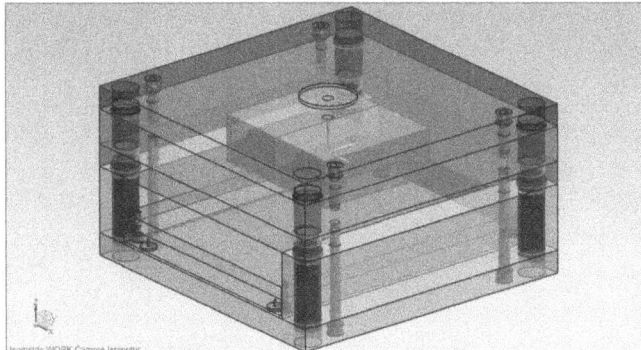

Figure 6-22 *The mold assembly*

Creating the Gate

In this section, you will create the gate.

1. Choose the **Design Fill** tool from the **Main** gallery of the **Mold Wizard** tab; the
    Design Fill dialog box is displayed with the **Information** window, refer to Figure 6-23.

2. Specify the gate type, Fan gate, from the **Member Select** panel in the **Reuse Library**.

3. Specify the parameters for the gate in the **Details** rollout of the dialog box. Make
    sure value for **Section_Type** is **Circular** and **D** is **3**. Click on the **Select Object** area
    in the **Placement** rollout and specify the position and orientation of the gate, refer to
    Figure 6-24. Repeat the steps for placing other gates.

4. After placing the gates, choose the **OK** button to close the dialog box.

> **Tip**
> *For better visibility while positioning the gate, you can hide solid bodies.*

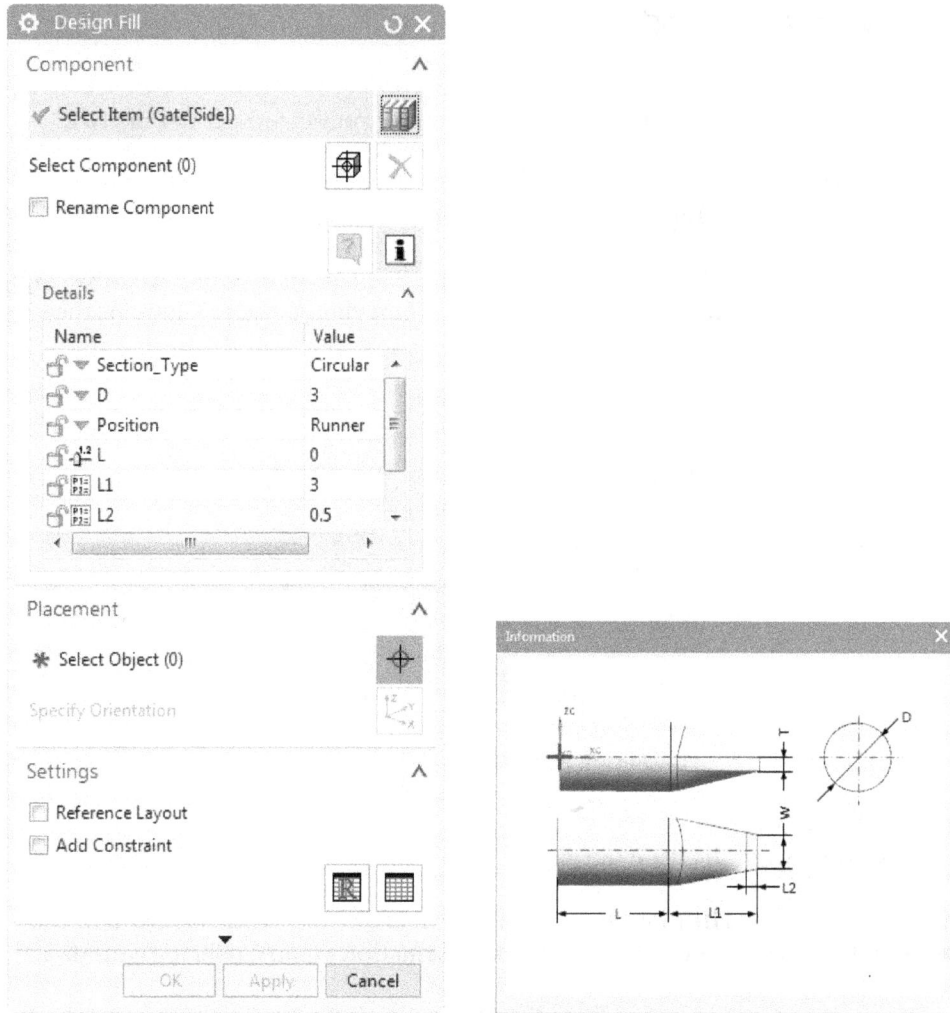

Figure 6-23 The **Design Fill** *dialog box and the* **Information** *window*

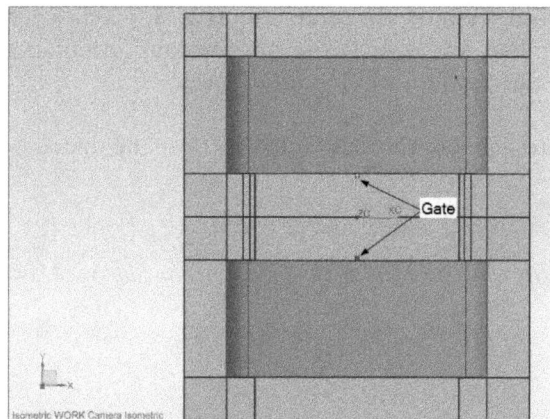

Figure 6-24 The position and orientation of the gate in the model

Creating the Runner

In this section, you will create the runner.

1. Choose the **Design Fill** tool from the **Main** gallery of the **Mold Wizard** tab; the **Design Fill** dialog box is displayed.

2. Select **Runner[2]** from the **Member Select** panel in the **Reuse Library**.

3. Select the **Circular** and **3** options from the **Section_Type** and **D** drop-down lists respectively. Enter **45** in the **L** edit box in the **Details** sub-rollout of the dialog box. Select the **Specify Point** area in the **Placement** rollout and position the runner, refer to Figure 6-25.

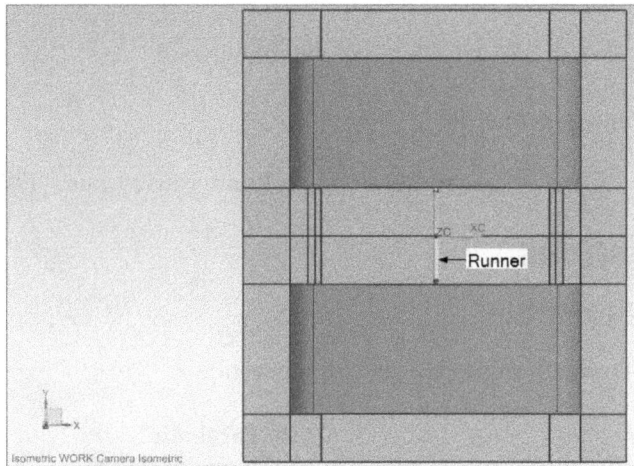

Figure 6-25 The position and orientation of runner in the model

4. After placing the runner, choose the **OK** button to close the dialog box.

Saving and Closing the File

1. Choose **Menu > File > Close > Save and Close** from the **Top Border Bar** to save and close the file.

Self-Evaluation Test

Answer the following questions and then compare them to those given at the end of this chapter:

1. Which of the following variables controls the diameter of the gate cross-section?

   (a) **D**   (b) **B**
   (c) **E**   (d) None of these

2. Which of the following variables controls the length of the gate?

   (a) **L, L1 and L2**   (b) **L, L2 and L3**
   (c) **L1, L2 and L3**   (d) None of these

3. Which of the following variables controls the cross-section shape of the gate?

    (a) **Section_Type**   (b) **Section**
    (c) **Area**   (d) None of these

4. Efficiency of a round runner is _____.

5. The _____ tool is used create the runner using sketch.

6. The _____ tool is used to create the gate.

7. The _____ gate is a curved variation of subsurface gate.

8. The _____ gate is also known as curved tunnel gate.

9. Efficiency of a Square runner is _____.

10. It is difficult to achieve precise dimension in a Runner Edge gate. (T/F)

Review Questions

Answer the following questions:

1. Which of the following is not a type of runner section ?

    (a) **Circular**   (b) **Parabolic**
    (c) **Decagon**   (d) None of these

2. On which of the following factors does the layout of a runner system depend ?

    (a) Number of impressions   (b) Shape of the components
    (c) Type of mold   (d) All of the above

3. Normally, the feed system comprises a sprue, _____, and _____.

4. Efficiency of a rectangular runner is _____.

5. Gate length is also known as _____.

6. The _____ button is used to position a gate.

7. A _____ gate is formed by machining a matching semi-circular channel.

8. Efficiency of a semi-circular runner is _____.

9. The _____ type of gate is used for tubular-type moldings when more than one impression is required in a simple two-plate mold.

10. Gate is a channel which connects the runner with a sprue bush. (T/F)

EXERCISES

Exercise 1

In this exercise, you will open the output file (exr_01_top_###) of Exercise 1 of Chapter 5. Create a gate and runner for the model shown in Figure 6-26 and save it.   **(Expected time: 1 hr)**

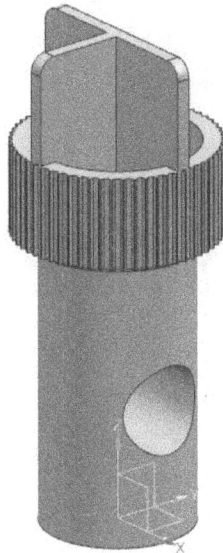

Figure 6-26 Model for Exercise 1

Exercise 2

In this exercise, you will open the output file (exr_02_top_###) of Exercise 2 of Chapter 5. Create a gate and runner for the model shown in Figure 6-27 and save it.   **(Expected time: 1 hr)**

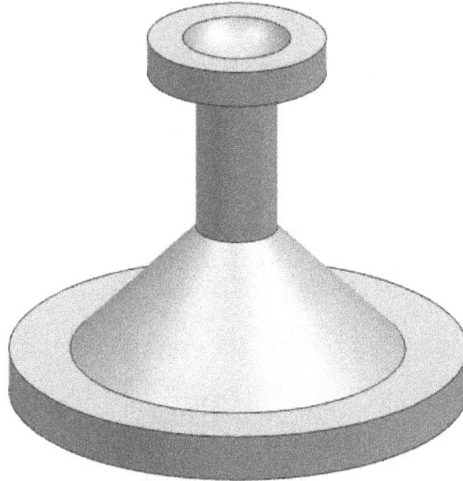

Figure 6-27 Model for Exercise 2

Chapter 7

Creating Sliders and Lifters

Learning Objectives

After completing this chapter, you will be able to:

- *Define undercut*
- *Define types of undercut*
- *Understand removal techniques of undercut*
- *Create the slider*
- *Create the lifter*

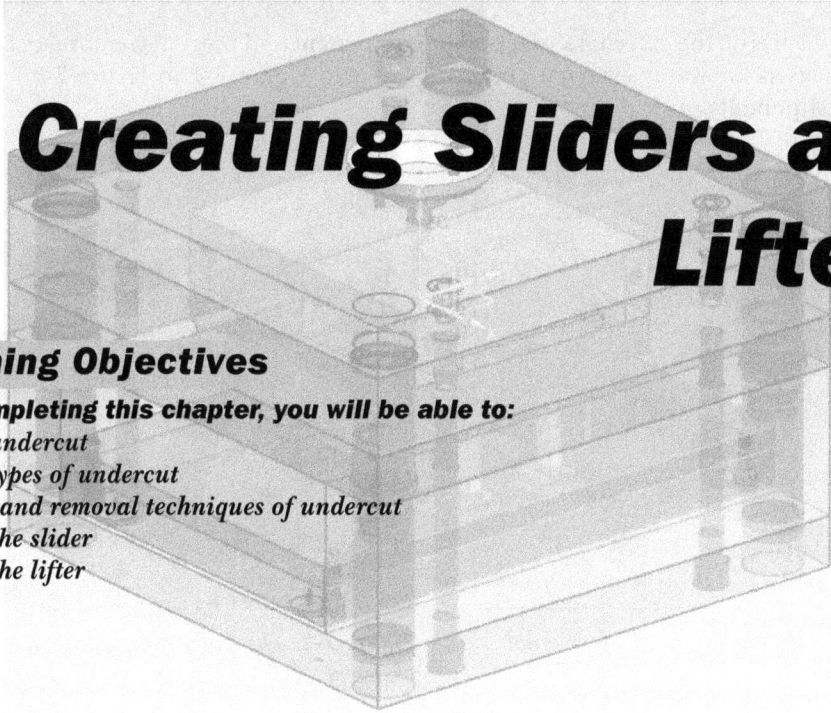

INTRODUCTION

In the previous chapter, you created gate, runner, and layout for the mold. In this chapter, you will learn to add sliders and lifters using various tools.

UNDERCUTS

Producing undercut parts in a mold is a very difficult task for the mold designers. Undercut is a recess or projection in the molding part. In undercut parts, mold designers should plan how to eject the part.

Any recess which is on the external surface of the component and prevents removal of component from the cavity is known as external undercut. Refer to Figure 7-1 to Figure 7-3 for external undercut components.

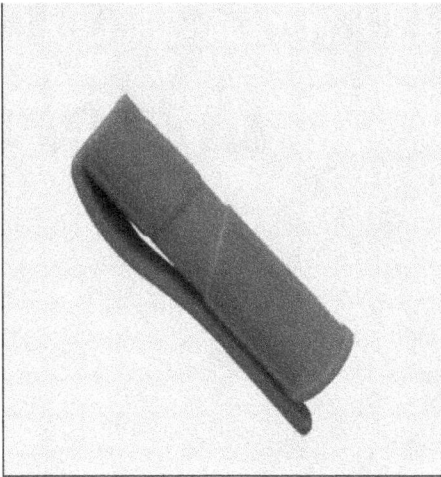

Figure 7-1 The pen cap with external undercut

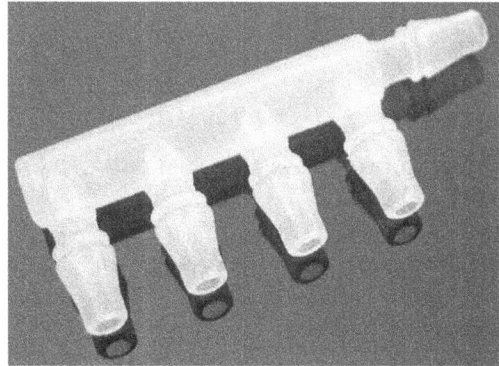

Figure 7-2 The water connector with external undercut

Figure 7-3 The threaded adapter with external undercut

A restriction which prevents the removal of any component from the core in line of draw is known as internal undercut. Refer to Figure 7-4 for internal undercut components.

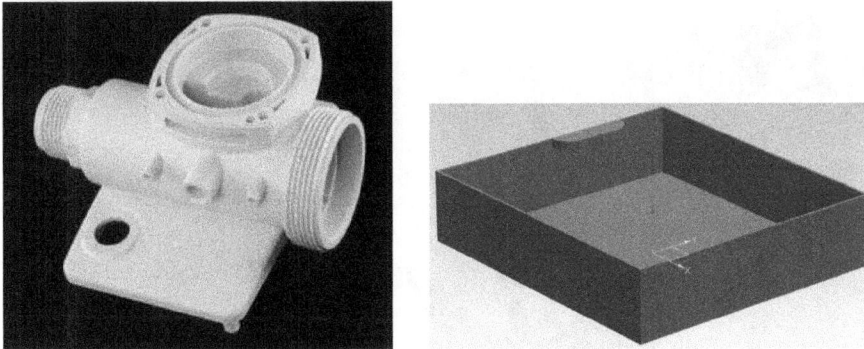

Figure 7-4 *Components with internal undercut*

TECHNIQUES USED FOR REMOVING UNDERCUTS

Several techniques are used for the removal of external and internal undercut of the components. Some of these techniques are mentioned next.

Techniques used for the removal of external undercuts:

(i) Use of mechanical pins (Finger Cam Actuation, Dog-Leg Cam Actuation)
(ii) Use of hydraulic and pneumatic cylinders

Techniques used for the removal of internal undercuts:

(i)   Force ejection
(ii) Use of collapsible core
(iii) Use of lifters
(iv) Use of unscrewing mechanisms (for threads)

External Undercut Actuation Techniques

External undercut techniques are used for removing external undercut of the component. Some of these techniques are mentioned next.

Finger Cam Actuation Technique

In this actuation technique, a hardened circular steel pin called finger pin is mounted at an angle in a fixed mold plate. Some splits are mounted on guide plates on the flat mold plate which has angled circular holes to accommodate these finger pins. Refer to Figures 7-5 and 7-6 for a slider assembly and its various components.

Figure 7-5 The different parts of a slider

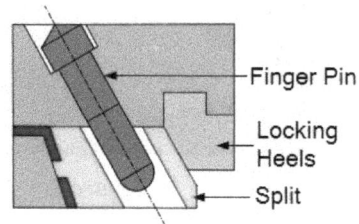

Figure 7-6 The slider assembly

Working of Sliders

Before doing calculation for split movement, you need to understand the working of slider. Figures 7-7 to 7-9 show the working of the slider. Figure 7-7 shows the slider in closed position. As the mold opens, the finger pin forces the split to move outward, as shown in Figure 7-8. Continuous movement of the moving half causes the molding to get ejected, as shown in Figure 7-9.

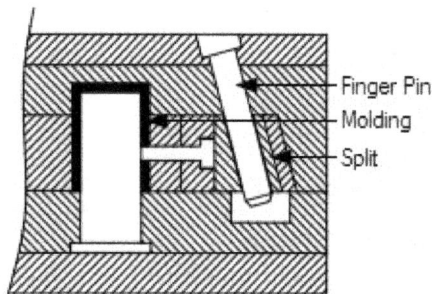

Figure 7-7 Slider in closed position

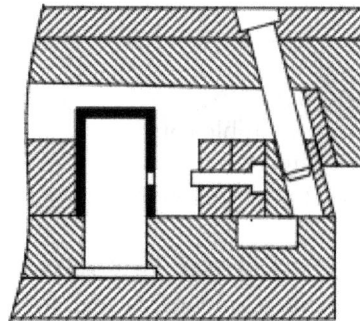

Figure 7-8 Intermediate position

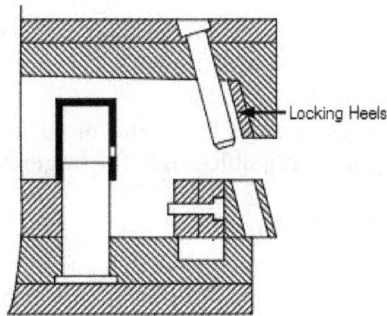

Figure 7-9 Fully open position of slider

Calculation for Split Movement

The distance traversed by the split across the face of mold plate is determined by the length and angle of the finger cam. The required movement of the split is dependent on the amount of component undercut. While specifying the split movement, designer should keep the split movement to minimum but ensure that the molded part can be easily and quickly removed from the mold. Equations (i) and (ii) given below represent the relation between split movement, angle of finger cam, working length of finger cam, and clearance. Figure 7-10 shows the split movement calculation.

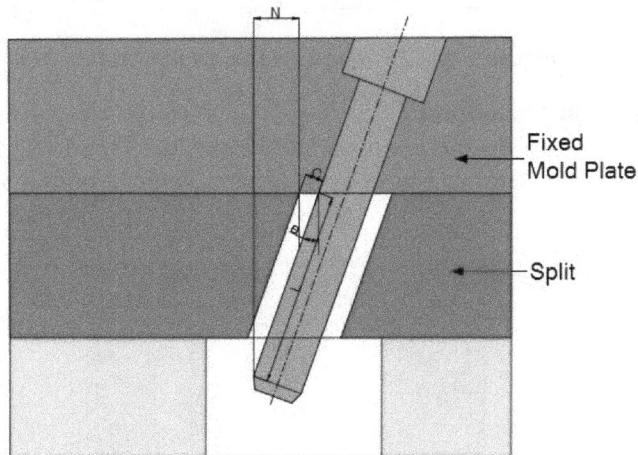

Figure 7-10 The split movement calculation

$$N = (L\sin) - (C/\cos) \qquad\qquad \text{-------(i)}$$

$$L = (N/\sin) + (2C/\sin2) \qquad\qquad \text{-------(ii)}$$

Where

N = Split movement

= Angle of finger cam

$$L = \text{Working length of finger cam}$$
$$C = \text{Clearance}$$

Suitable angle for finger cam is 10 degrees but if the mold height needs to be increased to accomodate large finger cam then permissible angle for finger cam should be 25 degrees.

Dog-Leg Cam Actuation Technique

This actuation system is used where a longer split delay time is required compared to finger cam actuation method. Designer should ensure that molding completely withdraws from the fixed half before retracting the side core. Figure 7-11 shows the Dog-Leg cam actuated mold.

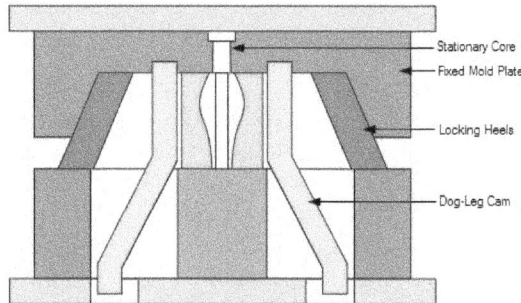

Figure 7-11 *The Dog-Leg cam actuated mold*

Calculation for Split Movement

The formula for calculating the opening movement, the length of cam, and the delay period are given below. Refer to Figure 7-12 for parameters required to calculate the split movement.

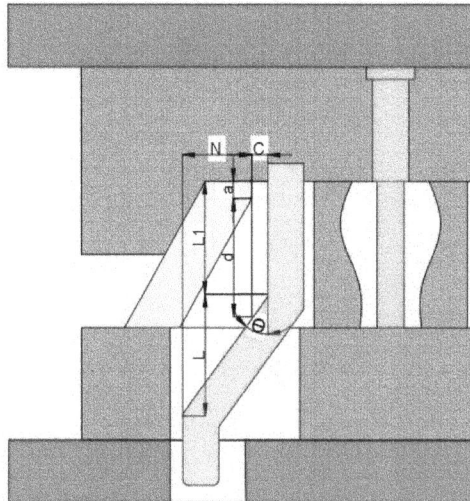

Figure 7-12 *The Split movement calculation*

$$N = L\tan - C \qquad\qquad \text{-------(i)}$$

$$d = (L1-a) + (C/\tan) \qquad\qquad \text{-------(ii)}$$

where

N	= movement of each split
L	= angled length of cam
L1	= straight length of cam
	= cam angle
C	= clearance
d	= delay
a	= length of straight portion of the hole

CREATING SLIDER

Ribbon: Mold Wizard > Main > Slide and Lifter Library

The **Slide and Lifter Library** tool helps you to create a slider in a mold. The slider helps to release external undercut in the given part. To create a slider, choose the **Slide and Lifter Library** tool from the **Main** gallery of the **Mold Wizard** tab; the **Slide and Lifter Design** dialog box will be displayed along with the **Information** window, as shown in Figures 7-13 and 7-14. Also, the **Reuse Library** navigator will be displayed on the left of the window. By default, the **Slide** folder is selected from the **SLIDE_LIFT** folder in the **MW Slide and Lifter Library** in the **Main** panel. Select the type of slider from the **Member Select** panel of the **Reuse Library** and specify the parameters in the **Details** rollout of the **Slide and Lifter Design** dialog box. Choose the **OK** button to close the dialog box.

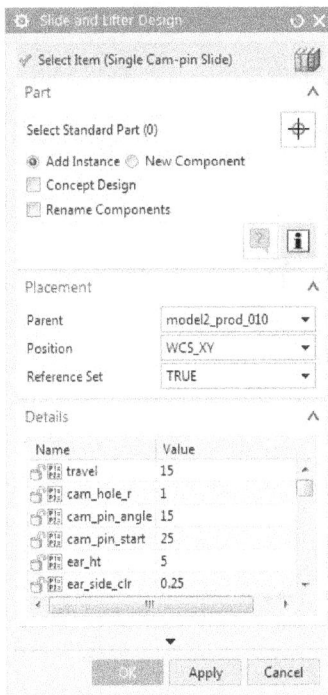

Figure 7-13 The Slide and Lifter Design dialog box

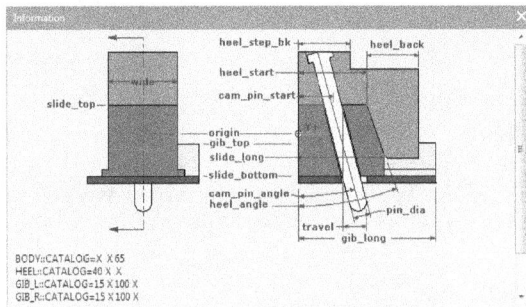

Figure 7-14 The Information window

CREATING LIFTERS

Ribbon: Mold Wizard > Main > Slide and Lifter Library

The **Slide and Lifter Library** tool also helps you to create a lifter in a mold. To create a lifter, choose the **Slide and Lifter Library** tool from the **Main** gallery of the **Mold Wizard** tab; the **Slide and Lifter Design** dialog box and the **Information** window will be displayed, refer to Figures 7-15 and 7-16. Select **Lifter** from the **SLIDE_LIFT** sub-tree of the **MW Slide and Lifter Library** tree in the **Reuse Library**. Select the type of lifter from the **Member Select** panel and specify the parameters in the **Details** rollout of the **Slide and Lifter Design** dialog box. Choose the **OK** button to close the dialog box.

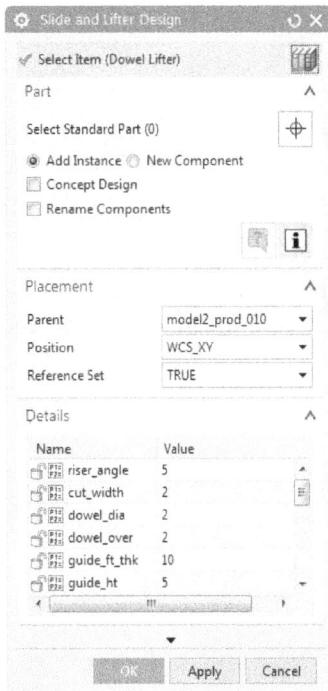

*Figure 7-15 The **Slide and Lifter Design** dialog box*

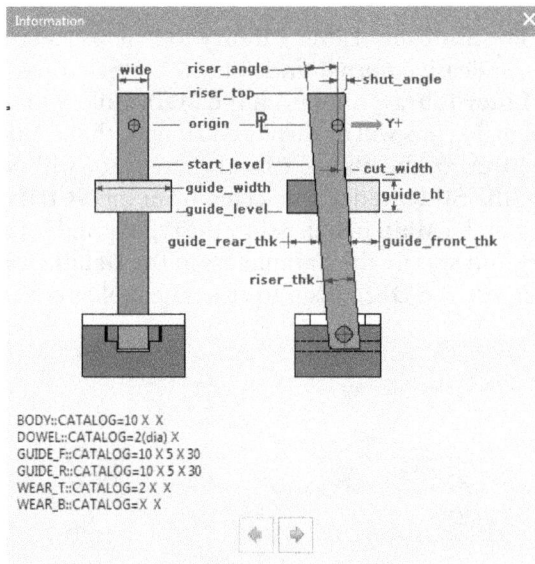

*Figure 7-16 The **Information** window*

CREATING BOUNDING BODY

Ribbon: Mold Wizard > Mold Tools > Bounding Body

The **Bounding Body** tool is used to create a box or a cylinder by selecting faces, solid bodies, facet bodies, edges and curves. This tool is used where the component has special areas such as a hole for moving a slider. To create a box, choose the **Bounding Body** tool from the **Mold Tools** gallery of the **Mold Wizard** tab; the **Bounding Body** dialog box will be displayed, refer to Figure 7-17. The options in this dialog box are discussed next.

Type Rollout

The options of this rollout are discussed next.

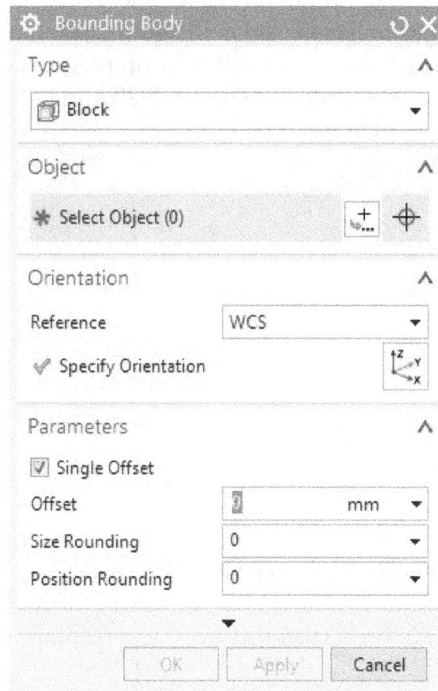

*Figure 7-17 The **Bounding Body** dialog box*

Block
This option helps you to create an associative block enclosing selected faces, edges, curves, solid bodies and facet bodies. This block can be aligned with specified orientation and its size can be varied with specified offset.

Center and Lengths
This option helps you to create a box that is centered at a selected point and positioned to the specified orientation. You can specify different values for length, width, and height.

Cylinder
This option helps you to create an associative cylinder enclosing selected faces, edges, curves, solid bodies, and facet bodies. The cylinder is aligned with the specified vector.

CREATING SPLIT BODIES

Ribbon: Mold Wizard > Mold Tools > Split Body

The **Split Body** tool is used to divide a solid body or sheet body into multiple bodies using a single face, a set of faces, or a datum plane. This tool helps you to create an associative split body feature that appears in the history of the model. You can update, edit or delete the feature. To create a split body, choose the **Split Body** tool from the **Mold Tools** gallery of the **Mold Wizard** tab; the **Split Body** dialog box will be displayed, refer to Figure 7-18. Also, you will be prompted to select target bodies to split. Select the solid body, refer to Figure 7-19. By default, the **Face or Plane** option is selected in the **Tool Option** drop-down list in **Tool** rollout. Next, choose the **Face or Plane** button; you will be prompted to select the tool faces or datum planes to split with. Select the plane, refer to Figure 7-20 and then choose the **OK** button; the component gets splitted, refer to Figure 7-21.

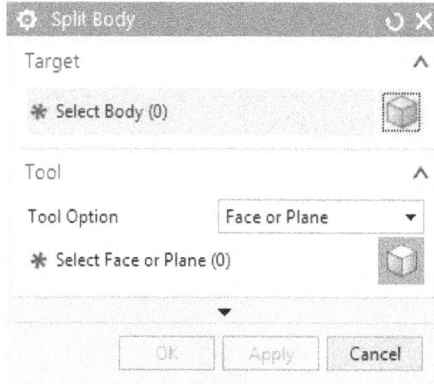

*Figure 7-18 The **Split Body** dialog box*

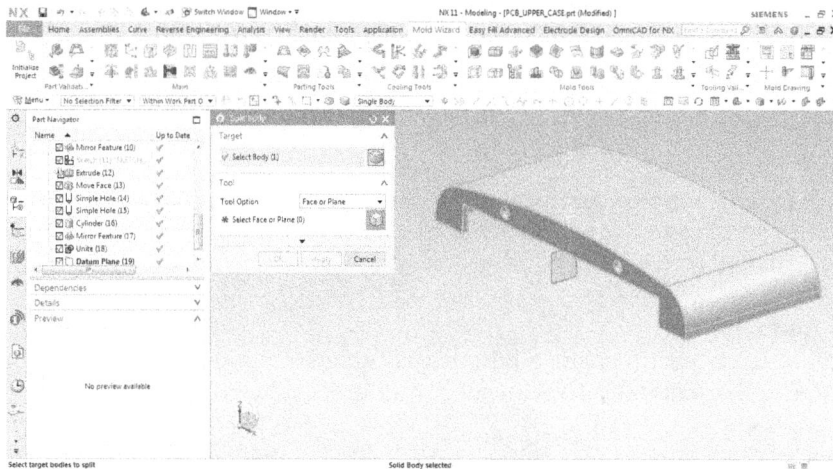

Figure 7-19 The component selected for splitting the body

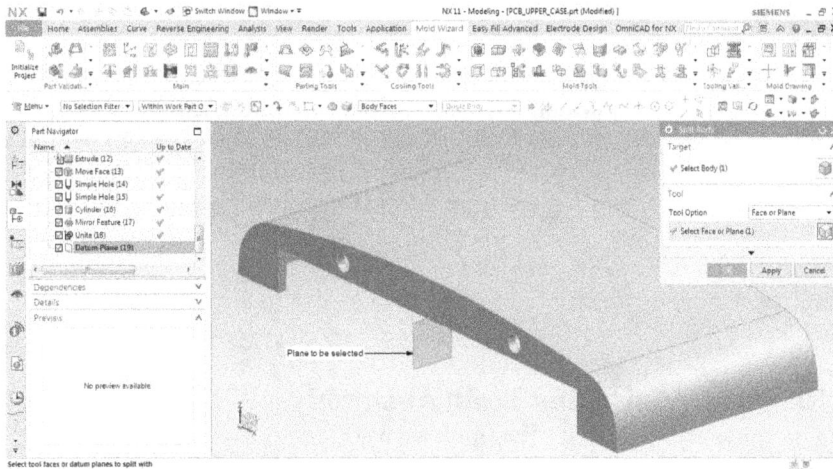

Figure 7-20 *The face selected for splitting the body*

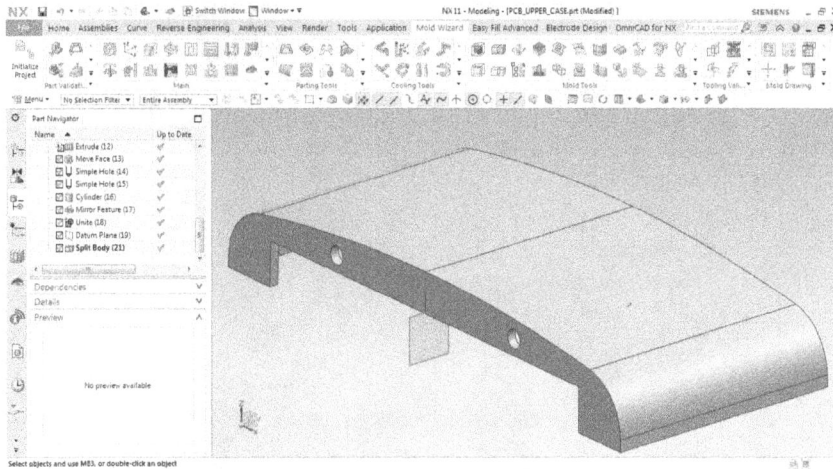

Figure 7-21 *The resulting split body*

TUTORIAL

To perform the tutorial, you need to download the zipped file named as *c07_NX_11.0_input* from the **Input Files** section of the CADCIM website. The complete path for downloading the file is:

Textbooks > CAD/CAM > NX_Mold > Mold Design using NX 11.0: A Tutorial Approach > Input Files

After the file is downloaded, extract the folder. In this folder, you will find Tut1 folder containing input file for Tutorial 1.

Tutorial 1

In this tutorial, you will add the lifter to the model (PCB_UPPER_CASE_top_###) contained in Tut 1 folder that you have downloaded. After adding the lifter, save the file.

(Expected time: 2 hr)

The following steps are required to complete this tutorial:

a.   Start NX and open the model, refer to Figure 7-22.
b.   Add lifter, refer to Figure 7-26.
c.   Save the file.

Starting NX and Opening the Mold Assembly

First, you need to start NX and then open a saved file.

1.   Double-click on the shortcut icon of NX on the desktop of your computer to start NX.

2.   Choose the **Open** button from the **Standard** group of the **Home** tab or choose **Menu > File > Open** from the **Top Border Bar**; the **Open** dialog box is displayed.

3.   Browse to **PCB_UPPER_CASE_top_###** in Tut 1 folder; the **PCB_UPPER_CASE_top_###** is displayed in the **File name** drop-down list. Next, choose the **OK** button; the assembly is displayed, refer to Figure 7-22.

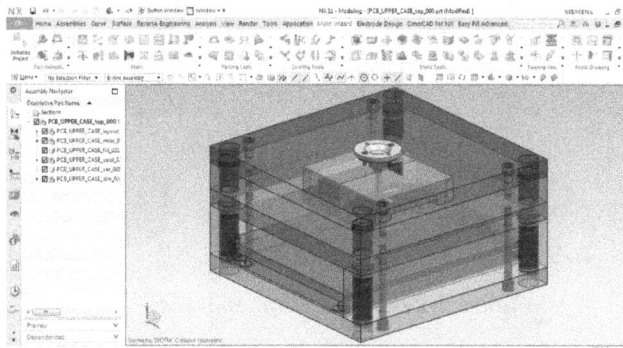

Figure 7-22 Mold assembly in Mold environment

> **Note**
> *When you open the file, the **Update event list** dialog box is displayed along with the **Information** dialog box. Choose the cancel and close button to close the dialog box.*

Adding Lifter to the Mold Assembly

In this section, you will add lifter to mold assembly.

1.   Choose the **Slide and Lifter Library** tool from the **Main** gallery of the **Mold Wizard** tab; the **Slide and Lifter Design** dialog box is displayed along with the **Information** window.

2. Choose the **Lifter** folder in the **Main Panel** and **Dowel Lifter** in the **Member Select** panel of the **Reuse Library**; the **Slide and Lifter Design** dialog box gets modified along with the **Information** window, refer to Figures 7-23 and 7-24.

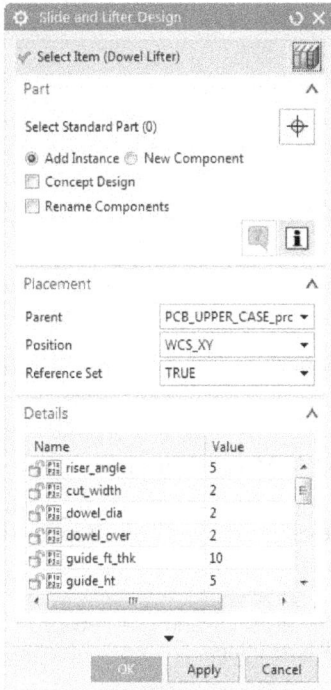

*Figure 7-23 The **Slide and Lifter Design** dialog box*

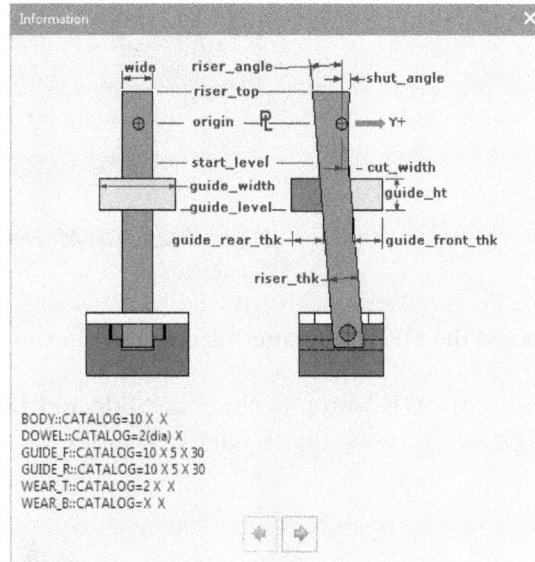

*Figure 7-24 The **Information** window*

3. Enter **25** in the **riser_top** edit box of the **Details** rollout and choose the **Apply** button.

   You can position the lifter to place it in a correct way.

4. Choose the **Select Standard Part** area of the **Part** rollout and then select the lifter which you want to place.

5. Choose the **Reposition** button in the **Part** rollout; the **Move Component** dialog box is displayed, refer to Figure 7-25.

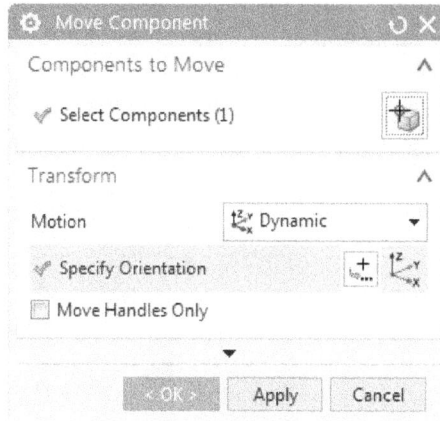

Figure 7-25 The Move Component dialog box

6. Choose the **OK** button after placing the lifter to close the **Move Component** dialog box.

7. Choose the **OK** button to close the **Slide and Lifter Design** dialog box. Repeat the same procedure for other lifters, refer to Figure 7-26 for arrangement of lifters.

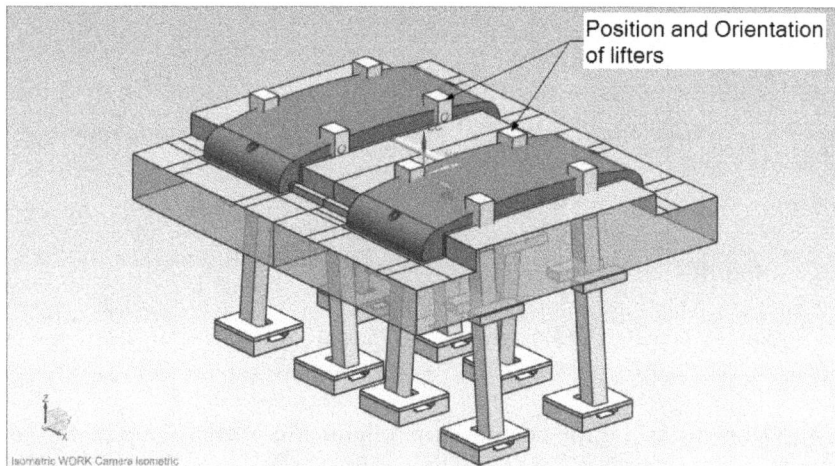

Figure 7-26 The arrangement of lifters

Trimming the Lifter in Mold Assembly

Now, use the **Trim Mold Components** tool to trim the lifters for mold assembly.

1. Choose the **Trim Mold Components** tool from the **Mold Tools** gallery; the **Trim Mold Components** dialog box is displayed, refer to Figure 7-27. Also, you are prompted to select target bodies. Select the lifters, refer to Figure 7-28.

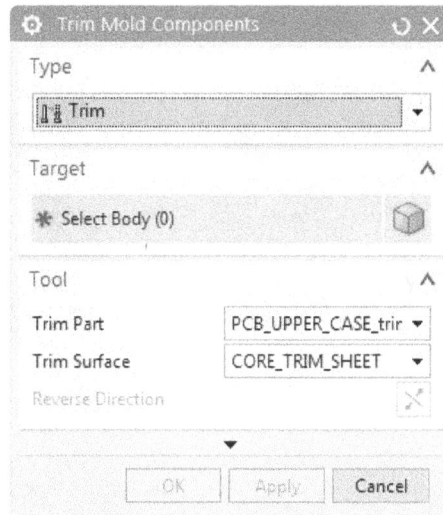

*Figure 7-27 The **Trim Mold Components** dialog box*

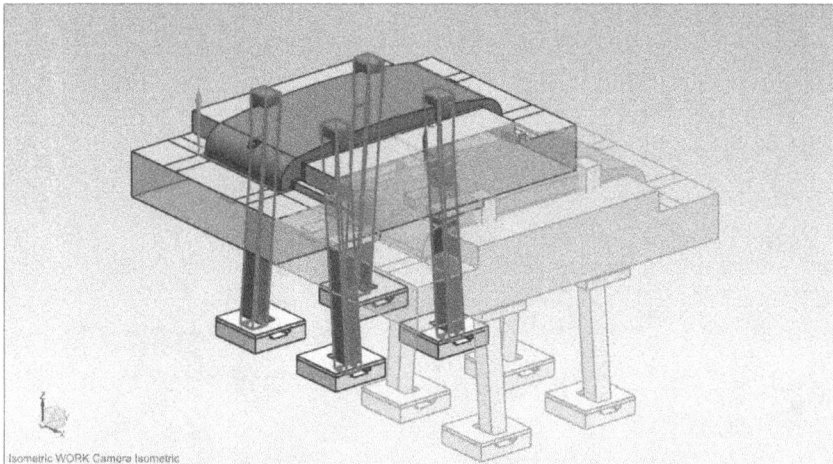

Figure 7-28 The lifters selected for trimming

By default, the **Trim** option is selected in the drop-down list in the **Type** rollout of the **Trim Mold Components** dialog box.

2. Choose the **OK** button to trim the lifters.

Pocketing in Mold Assembly

Now, you need to use the **Pocket** tool to create pockets.

1. Choose the **Pocket** tool from the **Main** gallery; the **Pocket** dialog box is displayed, refer to Figure 7-29. Also, you are prompted to select target bodies. Select the core, refer to Figure 7-30.

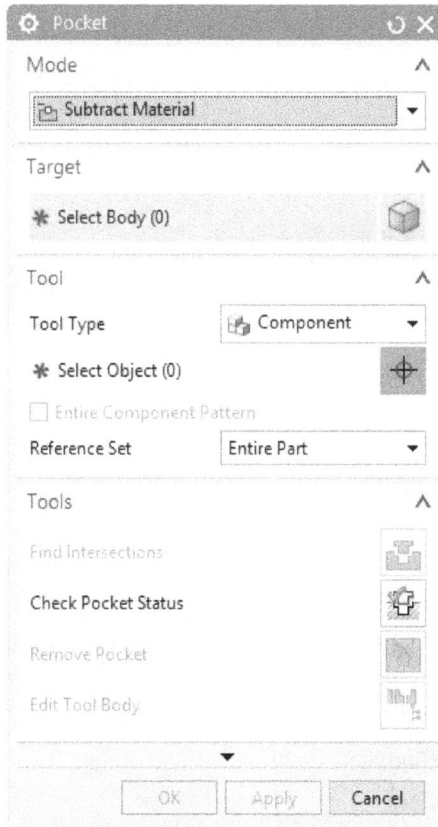

Figure 7-29 The **Pocket** dialog box

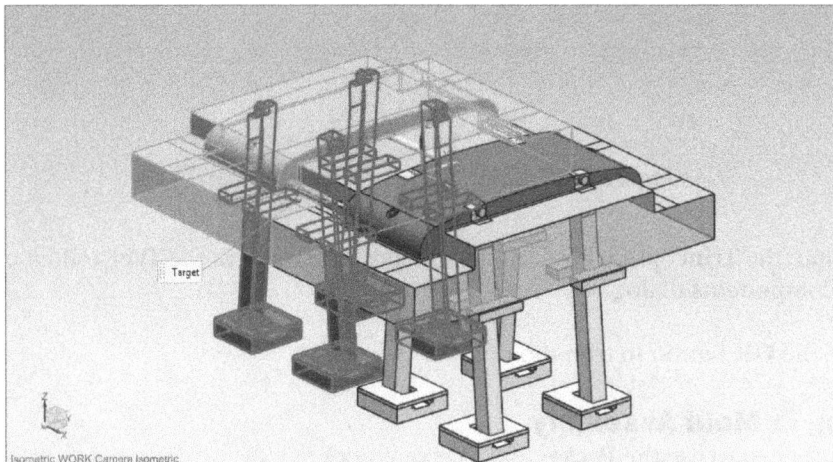

Figure 7-30 The lifters selected for pocketing

2. Click in the **Select Object** area in the **Tool** rollout; you are prompted to select the tool parts. Select the lifters and then choose the **OK** button. Refer to Figure 7-31 for lifters arrangement after pocket operation.

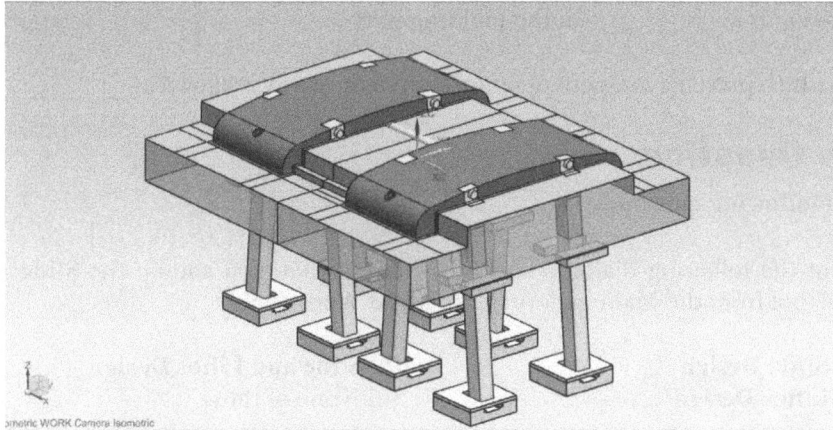

Figure 7-31 The lifters arrangement after pocketing

Saving and Closing the File

1. Choose **Menu > File > Close > Save and Close** from the **Top Border Bar** to save and close the file.

Self-Evaluation Test

Answer the following questions and then compare them to those given at the end of this chapter:

1. Which of the following techniques is used to remove external undercuts?

    (a) **Force ejection**   (b) **Collapsible core**
    (c) **Mechanical pins**   (d) None of these

2. Which of the following techniques is used to remove internal undercuts?

    (a) **Force ejection**   (b) **Collapsible core**
    (c) **Lifters**   (d) All of the above

3. Which of the following variables controls the riser angle of the lifter?

    (a) **riser_angle**   (b) **angle**
    (c) **A**   (d) None of these

4. Which of the following tools is used to create the lifter?

    (a) **Slide**   (b) **Slide and Lifter Library**
    (c) **Lifter**   (d) None of these

5. An undercut is a _____ in the molding part.

6. A recess which prevents component removal from the cavity is known as _____ undercut.

Review Questions

Answer the following questions:

1. Which of the following dialog boxes is displayed when you choose the **Slide and Lifter Library** tool from the **Main** gallery of the **Mold Wizard** tab?

   (a) **Slide Design**   (b) **Slide and Lifter Design**
   (c) **Lifter Design**   (d) None of these

2. Which actuation method is used when split delay is required during actuation?

   (a) **Finger cam**   (b) **Dog leg cam**
   (c) **Hydraulic cylinders**   (d) None of these

3. Which of the following is a suitable angle for finger cam?

   (a) **10°**   (b) **30°**
   (c) **45°**   (d) **60°**

4. Which of the following tools is used to trim a lifter?

   (a) **Trim**   (b) **Trim Mold**
   (c) **Trim Mold Components**   (d) None of these

5. Which of the following tools is used to create a pocket?

   (a) **Pocket**   (b) **Hole**
   (c) **Patch Surface**   (d) None of these

6. A recess which prevents component removal from the core is known as _____ undercut.

EXERCISES

Exercise 1

In this exercise, you will open the output file (exr_01_top_###) of Exercise 1 of Chapter 6. Next, you will add a slider in the mold base of the model shown in Figure 7-32 and save it.

(Expected time: 90 min)

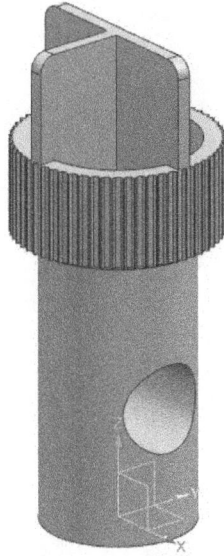

Figure 7-32 Model for Exercise 1

Exercise 2

In this exercise, you will open the output file (exr_02_top_###) of Exercise 2 of Chapter 6. Next, you will add a slider in the mold base of the model shown in Figure 7-33 and save it.

(Expected time: 90 min)

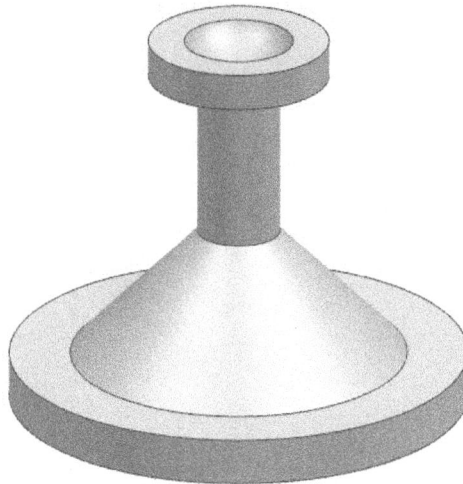

Figure 7-33 Model for Exercise 2

Answers to the Self-Evaluation Test

1. c, 2. d, **3.** a, **4.** b, **5.** recess, **6.** External

Chapter 8

Creating Ejection and Cooling Systems

Learning Objectives

After completing this chapter, you will be able to:

- *Understand the ejector grid*
- *Undertand the ejector plate assembly*
- *Understand the ejection techniques*
- *Understand the cooling process*

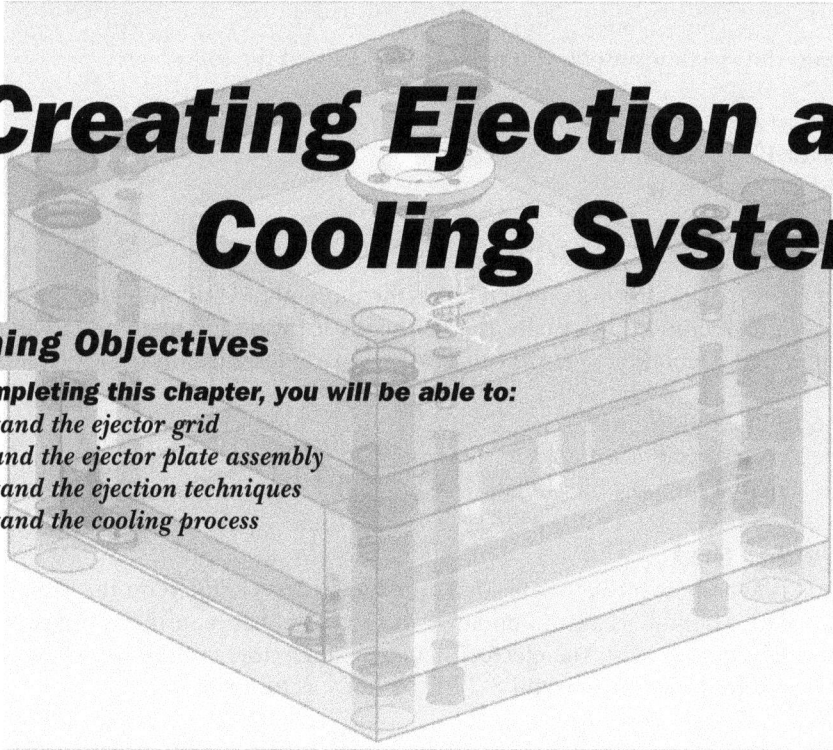

INTRODUCTION

In this chapter, you will learn to create ejection and cooling systems. The ejection system ejects the component from the mold and the cooling system cools the hot molten material.

EJECTION SYSTEM

When the molten material solidifies, it shrinks on the core. Due to shrinkage, it becomes difficult to remove the molding from the core. For removing the solidified material from the mold, an ejection system is required behind the moving platen to push out the solidified product from the core.

To understand the ejection system, you need to understand the following:

(i)   Ejector Grid
(ii) Ejector Plate Assembly
(iii) Ejection Techniques

Ejector Grid

Ejector Grid is that part of the mold which holds the mold plates and provides space for ejector plate assembly for operation. Normally, the grid consists of a back plate on to which support blocks are mounted. Generally, there are three alternative designs:

(i)   The In-line Ejector grid
(ii)   The Frame-type Ejector grid
(iii) The Circular Support block grid

In-line Ejector Grid

In this design, two support blocks are used. These blocks are mounted on the back plate of the grid. Ejector plate assembly is accommodated in the space between the two parallel support blocks. During ejection process, the ejector plate assembly moves in the space provided in the support blocks. Refer to Figure 8-1 and 8-2 for In-line ejector grid.

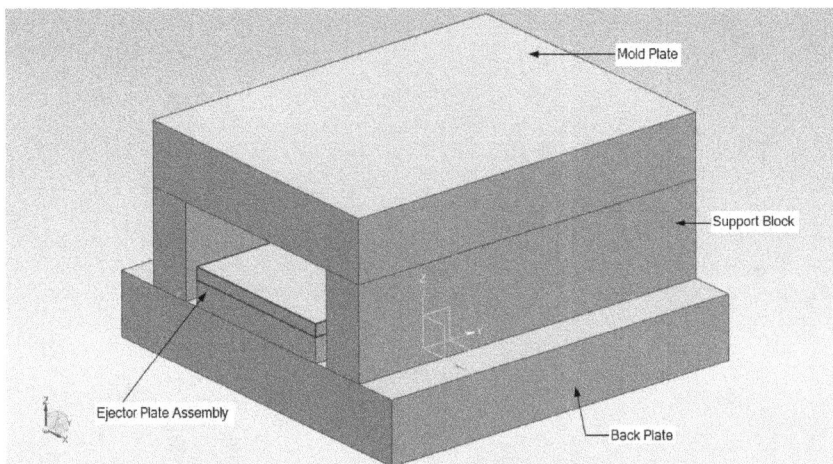

Figure 8-1 *The In-line ejector grid (isometric view)*

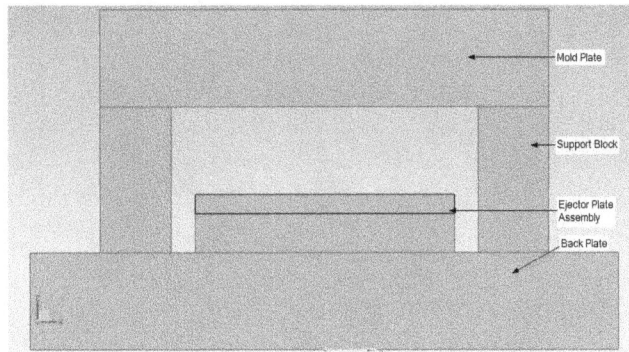

Figure 8-2 *The In-line ejector grid (side view)*

Frame Type Ejector Grid

Rectangular frame type ejector grid is the commonly-used frame type. It has four support blocks suitably mounted on the back plate of the grid. This design is preferred by mold designers due to the following reasons:

(i)   It is simple and cheap to manufacture
(ii) The ejector plate assembly is completely enclosed, thereby preventing foreign bodies entering the system

Refer to Figure 8-3 for the frame-type ejector grid.

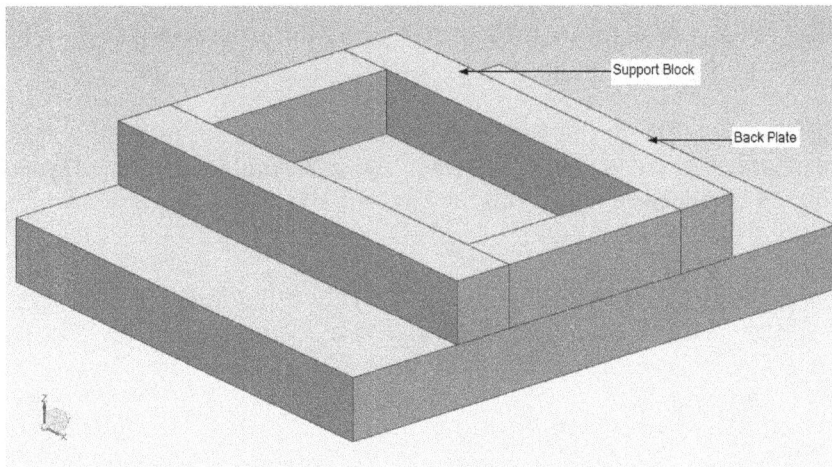

Figure 8-3 *The Frame type ejector grid*

Circular Support Block Grid

Circular support block grids are used for large molds. Refer to Figure 8-4 for the circular support block grid.

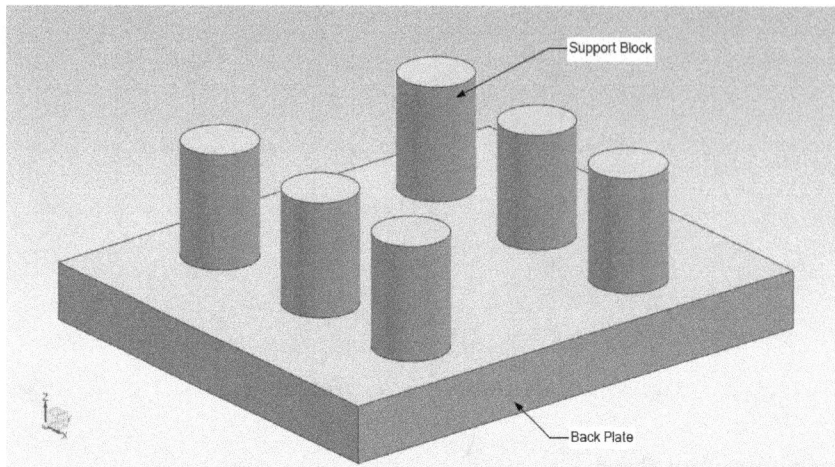

Figure 8-4 The circular support block grid

EJector Plate Assembly

Ejector plate assembly is an assembly comprising the ejector plate, retaining plate, and ejector rod.

Ejector Plate

It is a steel plate and is used to actuate the ejector system. This ejector plate receives direct actuated force from the injection machine.

The size of the ejector plate depends on the part shape or the profile, and the thickness depends upon the force required by the pin to strip the part.

Retaining Plate

It is a steel plate which is used to retain the ejector pin, sprue puller pin, guided ejector bushing, and return pins. The plate is screwed with the ejector plate by a socket-headed screw.

Ejector Rod

This rod is used to actuate the ejector plate.

Refer to Figure 8-5 for Ejector plate, Retaining plate, and Ejector rod assembly.

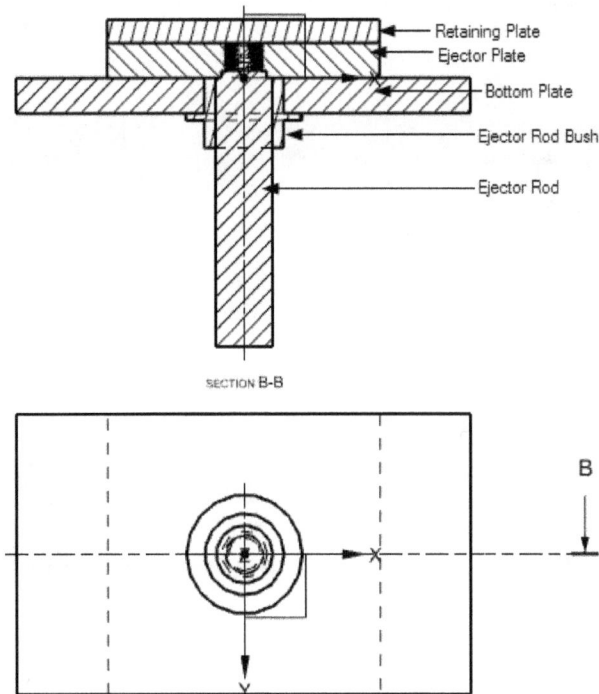

Figure 8-5 *The ejector plate, retaining plate and ejector rod assembly*

EJECTION TECHNIQUES

When a molded part cools, it shrinks depending on the material being processed. The choice of technique selected for the ejection process depends on the shape of the molding. The basic ejection techniques are:

(i)   Pin ejection
(ii) Sleeve ejection
(iii) Bar ejection
(iv) Blade ejection
(v)   Air ejection
(vi) Stripper plate ejection

Design Ejector Pin

This tool helps you create ejector pins used for ejection. To create an ejector pin, select **MW Standard Part Library** from the **Reuse Library**. Next, select the **Ejection** folder from the **DME_MM** folder and then double-click on **Ejector Pin(Shouldered)** in the **Member Select** rollout; the **Standard Part Management** dialog box will be displayed along with the **Information** window, refer to Figure 8-6 and 8-7. Place the ejection pin such that ejection takes place and then choose the **OK** button to close the dialog box, refer to Figure 8-8.

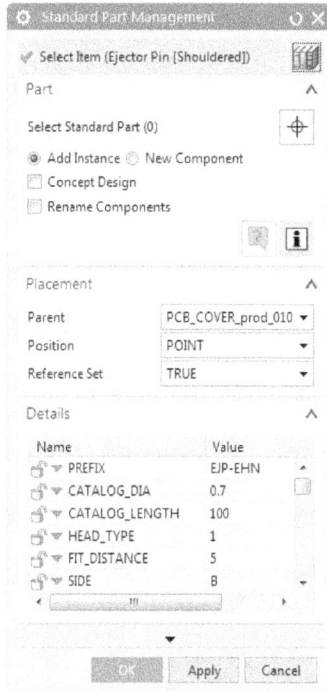

Figure 8-6 *The* **Standard Part Management** *dialog box*

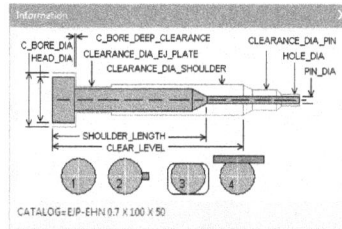

Figure 8-7 *The* **Information** *window*

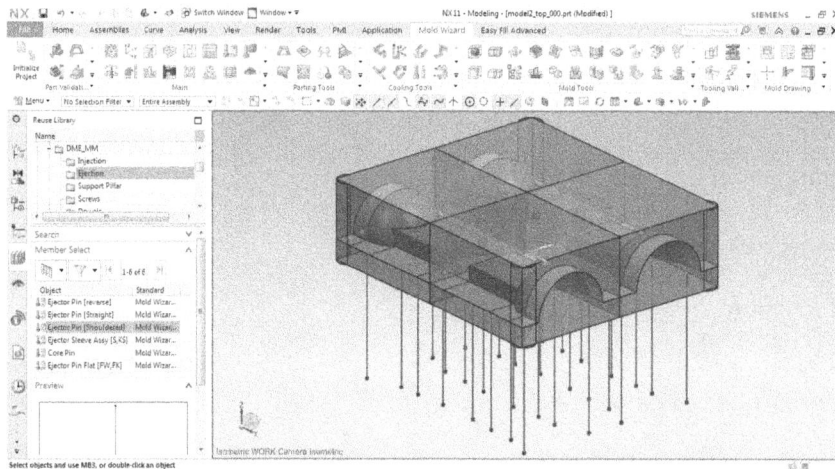

Figure 8-8 *The placement of ejector pin*

Ejector Pin Post Processing

Ribbon:	Mold Wizard > Main gallery > Ejector Pin Post Processing

This tool helps you trim the ejector pin. To trim the ejector pin, choose the **Ejector Pin Post Processing** tool from the **Main** gallery of the **Mold Wizard** tab; the **Ejector Pin Post**

Processing dialog box will be displayed, refer to Figure 8-9. Also, you will be prompted to select the target ejector pins. Select the ejector pins that you need to trim and then choose the **OK** button; the pins will be trimmed and the dialog box will be closed.

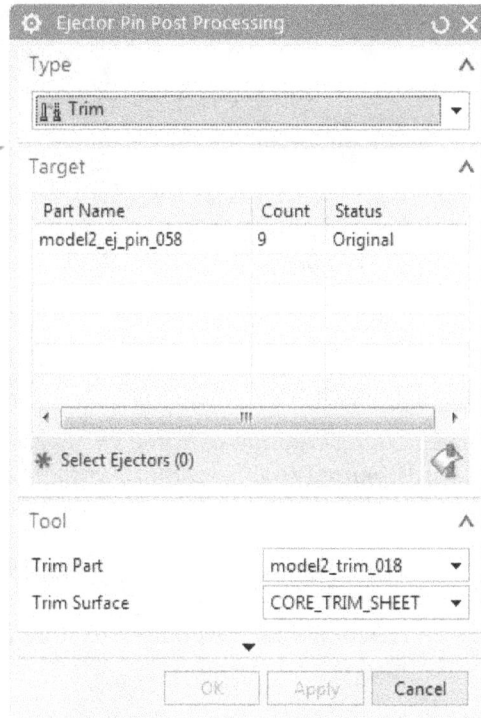

*Figure 8-9 The **Ejector Pin Post Processing** dialog box*

Cooling

Cooling is a process required to cool the component. The cycle time of cooling plays an important role in mold costing. NX provides you the **Cooling Tools** gallery in which all the tools related to cooling are available. The tools to create cooling channels are discussed next.

Pattern Channel

Ribbon: Mold Wizard > Cooling Tools gallery > Pattern Channel

This tool is used to create cooling channels using sketches or curves. To create a channel, choose the **Pattern Channel** tool from the **Cooling Tools** gallery of the **Mold Wizard** tab; the **Pattern Channel** dialog box will be displayed, refer to Figure 8-10. Also, you will be prompted to create sketch or select a section geometry. After creating the sketch, specify the channel diameter in the **Setting** rollout and then choose the **OK** button from the dialog box.

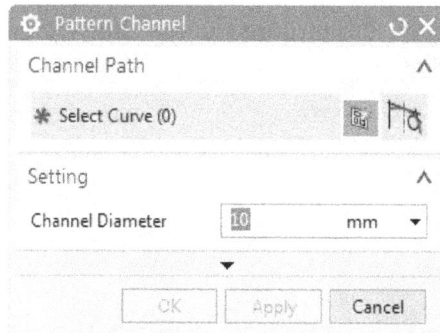

Figure 8-10   The **Pattern Channel** *dialog box*

Direct Channel

Ribbon: Mold Wizard > Cooling Tools gallery > Direct Channel

This tool is used to create cooling channel or baffle by specifying a point. To create a channel, choose the **Direct Channel** tool from the **Cooling Tools** gallery of the **Mold Wizard** tab; the **Direct Channel** dialog box will be displayed, refer to Figure 8-11. Also, you will be prompted to select an infer point. Select the start point of the channel. Next, select the **Distance** option from the **Motion** drop-down list of the **Channel Extrusion** rollout; you will be prompted to specify the vector direction. Specify the direction and then enter the length of the channel in the **Distance** edit box. Choose the **OK** button to close the dialog box.

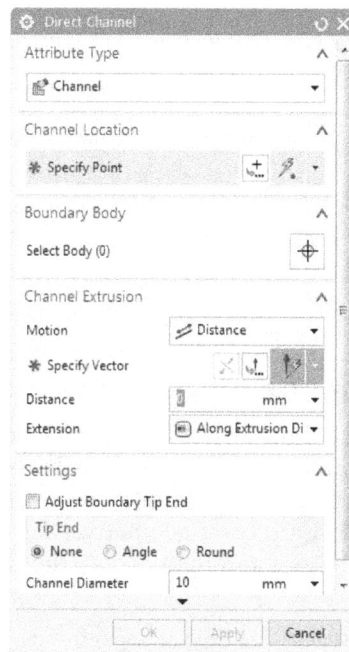

Figure 8-11   The **Direct Channel** *dialog box*

Extend Channel

Ribbon: Mold Wizard > Cooling Tools gallery > Extend Channel

This tool is used to extend the length of any cooling channel. To extend the length, choose the **Extend Channel** tool from the **Cooling Tools** gallery of the **Mold Wizard** tab; the **Extend Channel** dialog box will be displayed, refer to Figure 8-12, and you will be prompted to select the cooling channel. Select the cooling channel and specify the length of the cooling channel in the **Distance** edit box. Choose the **OK** button; the length will be extended and the dialog box will be closed.

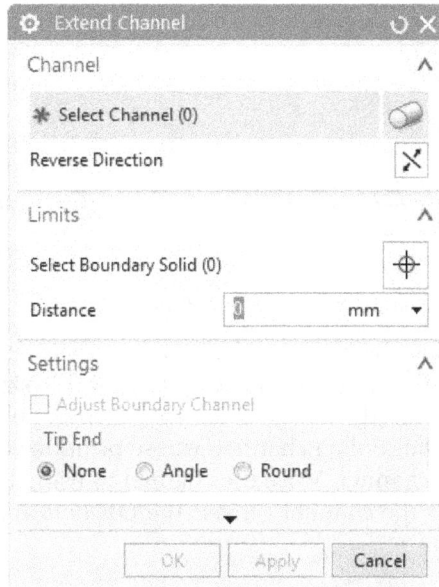

*Figure 8-12 The **Extend Channel** dialog box*

Connect Channels

Ribbon: Mold Wizard > Cooling Tools gallery > Connect Channels

This tool is used to connect the cooling channels. To do so, choose the **Connect Channels** tool from the **Cooling Tools** gallery of the **Mold Wizard** tab; the **Connect Channels** dialog box will be displayed, refer to Figure 8-13. Also, you will be prompted to select the first cooling channel to be connected. Select the first cooling channel; you will be prompted to select the second cooling channel. Select the second cooling channel and then choose the **OK** button to close the dialog box and the channels will be connected.

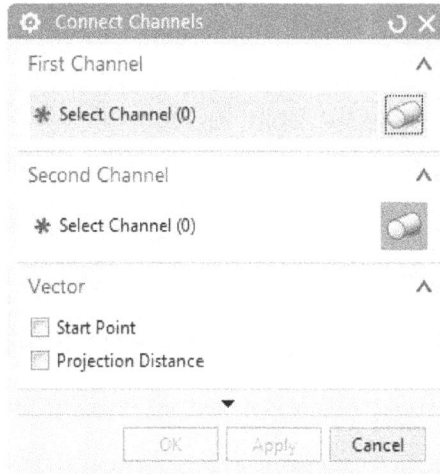

Figure 8-13 The **Connect Channels** *dialog box*

Adjust Channel

This tool is used to adjust the position of the cooling channel. To do so, choose the **Adjust Channel** tool from the **Cooling Tools** gallery of the **Mold Wizard** tab; the **Adjust Channel** dialog box will be displayed, refer to Figure 8-14. Also, you will be prompted to select the cooling channels. Select the cooling channels whose position you want to adjust; a dynamic triad will be displayed on the channel. Move the channel by using handles and angular handles of the triad. After positioning the channel, choose the **OK** button to close the dialog box.

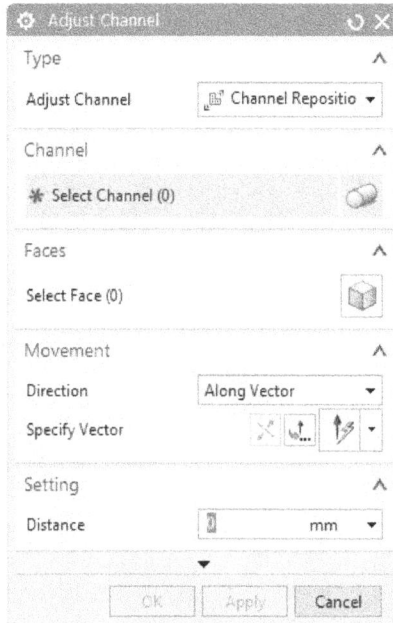

Figure 8-14 The **Adjust Channel** *dialog box*

Cooling Fittings

This tool is used to add cooling fitting components to the cooling channels. To add the components, choose the **Cooling Fittings** tool from the **Cooling Tools** gallery of the **Mold Wizard** tab; the **Cooling Fittings** dialog box will be displayed, refer to Figure 8-15, and you will be prompted to select the channel. Select the cooling channel and then specify the point. Next, choose the **OK** button; the dialog box will be closed and you will notice O-rings in the mold cooling channel at the specified point.

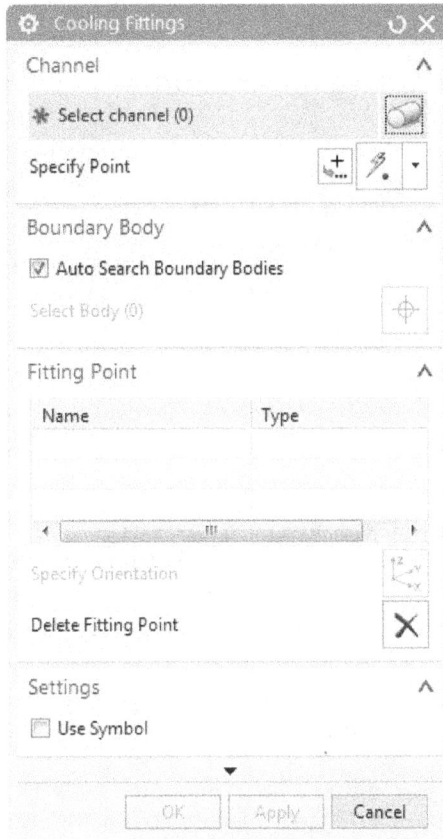

*Figure 8-15 The **Cooling Fittings** dialog box*

TUTORIALS

In these tutorials, you will use ejector pin and cooling channels for ejecting and cooling the given component. To perform the tutorials, you need to download the zipped file named as *c08_NX_Mold_input* from the **Input Files** section of the CADCIM website.
The complete path for downloading the file is:

> *Textbooks > CAD/CAM > NX_Mold > Mold Design using NX 11.0: A Tutorial Approach > Input Files*

After the file is downloaded, extract the folder. In this folder, you will find Tut1 and Tut2 folders containing input files for Tutorial 1 and Tutorial 2.

Tutorial 1

In this tutorial, you will add the ejector pin and cooling channel to the model (PCB_COVER_top_###) contained in Tut 1 folder that you have downloaded. After adding the ejector pin and cooling channel, save the file.   **(Expected time: 2 hr)**

The following steps are required to complete this tutorial:

a.   Start NX and open the file, refer to Figure 8-16.
b.   Add ejector pin and cooling channel, refer to Figures 8-19 and 8-24.
c.   Save the file.

Starting NX and Opening the Mold Assembly

First, you need to start NX and then open a saved file.

1.   Double-click on the NX shortcut icon on the desktop of your computer.

2.   Choose the **Open** button from the **Standard** group of the **Home** tab or choose **Menu > File > Open** from the **Top Border Bar**; the **Open** dialog box is displayed.

3.   Browse to **PCB_COVER_top_###** in Tut 1 folder; the **PCB_COVER_top_###** is displayed in the **File name** drop-down list. Then choose the **OK** button; the assembly is displayed, refer to Figure 8-16.

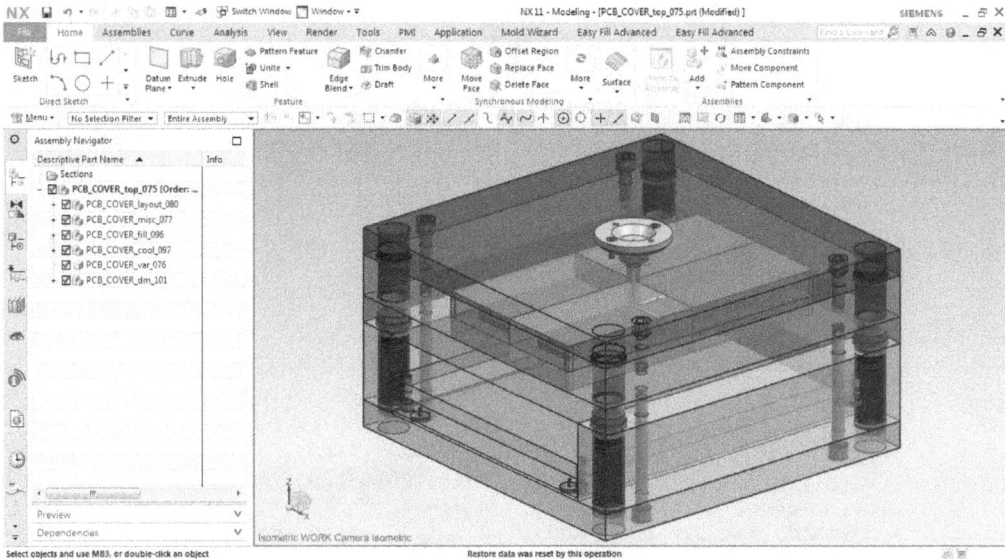

Figure 8-16 Mold assembly in Mold environment

Note
When you open the file, the Update event list dialog box is displayed along with the Information dialog box. Choose the cancel and close button to close the dialog box.

Adding Ejector Pin to the Mold Assembly

Now, you will add ejector pin to the mold assembly.

1. Choose **MW Standard Part Library** from the **Reuse Library**. Expand **MW Standard Part Library** and then choose the type of ejector pin from the **Ejection** folder of the **DME_MM** standard.

2. Double-click on **Ejector Pin[Shouldered]** in the **Member Select** panel; the **Standard Part Management** dialog box is displayed with the **Information** window, refer to Figure 8-17 (a) and (b).

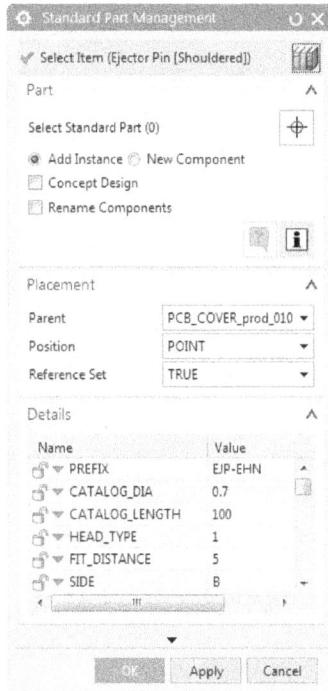

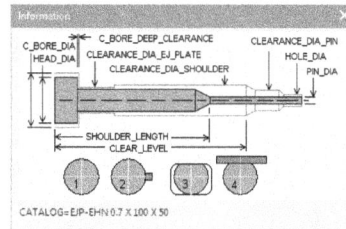

Figure 8-17 (a) *The **Standard Part**   **Figure 8-17 (b)** *The **Information** window*
Management *dialog box*

3.   Select **EJP-EHN**, **0.7**, **125**, and **1** from the **PREFIX**, **CATALOG_DIA**, **CATALOG_LENGTH**, and **HEAD_TYPE** drop-down lists respectively in the **Details** rollout. Choose the **Apply** button; the **Point** dialog box is displayed, refer to Figure 8-18. Specify the position of the ejector pin by specifying a point. Choose the **Cancel** button to close the dialog box. Next, choose the **OK** button to close the **Standard Part Management** dialog box and the **Information** window. Refer to Figure 8-19 for the arrangement of ejector pin. Refer to Table 8-1 for the positions of ejector pin.

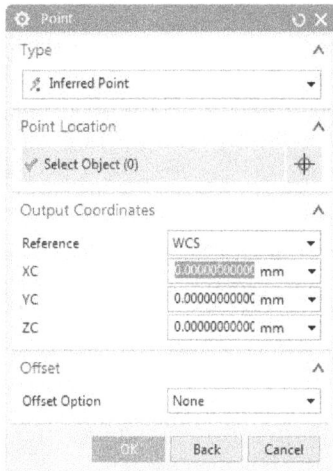

Figure 8-18 The **Point** dialog box

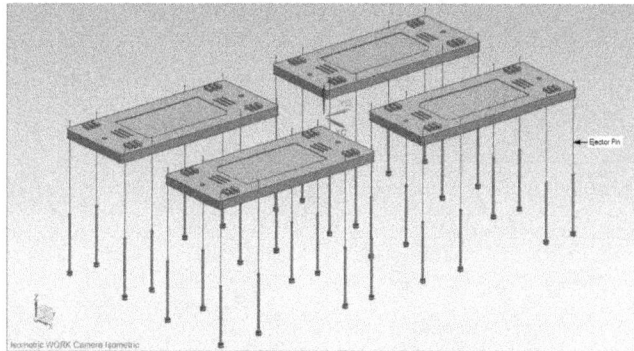

Figure 8-19 The arrangement of ejector pin in the mold assembly

Table 8-1 Positions of ejector pin

Ejector Pin	XC	YC
1	-27	-175
2	-27	-30
3	-27	-160
4	-27	-45
5	-92	-175
6	-92	-160
7	-92	-30
8	-92	-45

Trimming the Ejector Pin

You will trim the ejector pin in the mold assembly.

1.   Choose the **Ejector Pin Post Processing** tool from the **Main** gallery of the **Mold Wizard** tab; the **Ejector Pin Post Processing** dialog box is displayed, refer to Figure 8-20.

2.   Select the ejector pins which need to be trimmed.

3.   Choose the **Apply** button to trim the ejector pin; the **Information** window is displayed. Choose the **Close** button to close the window. Choose the **Cancel** button to close the dialog box. Refer to Figure 8-21 for the arrangement of ejector pin after trimming.

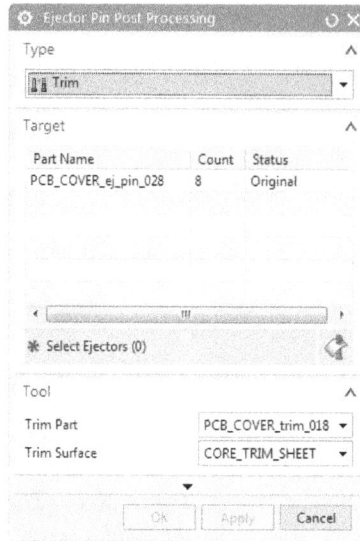

Figure 8-20 The *Ejector Pin Post Processing* dialog box

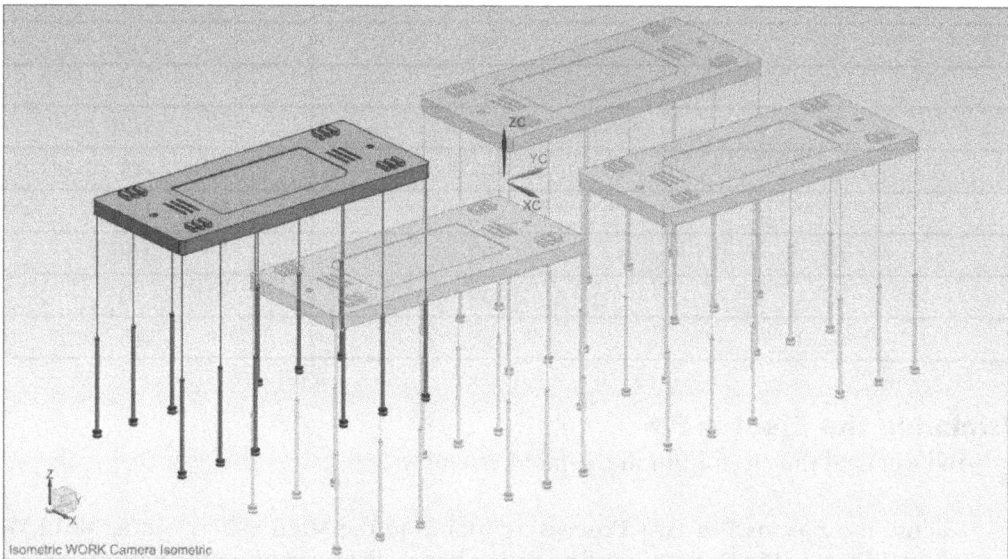

Figure 8-21 The arrangement of ejector pin after trimming process in the mold assembly

Adding the Cooling Channel

Now, you will add cooling channel to the mold component.

1. Choose the **Pattern Channel** tool from the **Cooling Tools** gallery of the **Mold Wizard** tab; the **Pattern Channel** dialog box is displayed, refer to Figure 8-22.

2. Choose the **Sketch Section** button in the **Channel Path** rollout; the **Create Sketch** dialog box is displayed, refer to Figure 8-23.

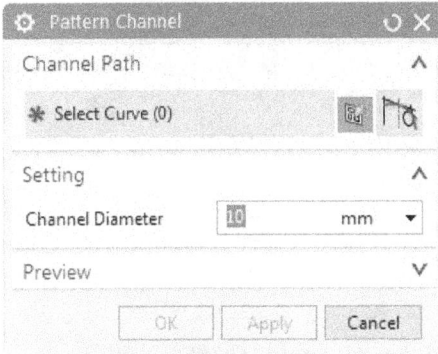

Figure 8-22 The Pattern Channel dialog box

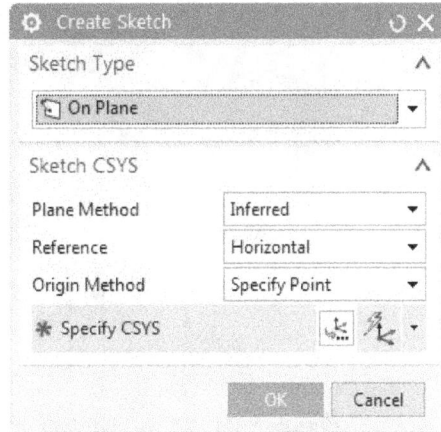

Figure 8-23 The Create Sketch dialog box

3. Create sketch for the cooling channel on a plane and then choose the **Finish** button to exit the sketch environment; the **Pattern Channel** dialog box is displayed. Enter **10** as the diameter of channel in the **Channel Diameter** edit box of the **Setting** rollout. Choose the **OK** button to close the dialog box. Refer to Figures 8-24 and 8-25 for understanding the position and orientation of the cooling channel.

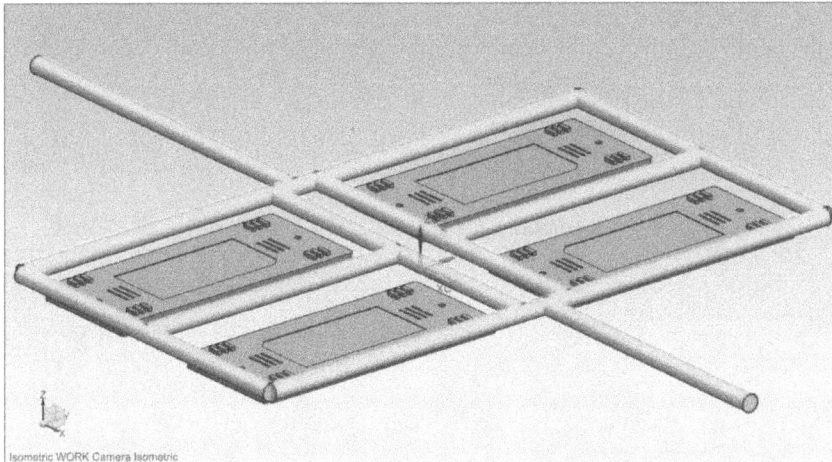

Figure 8-24 The orientation and position of cooling channel

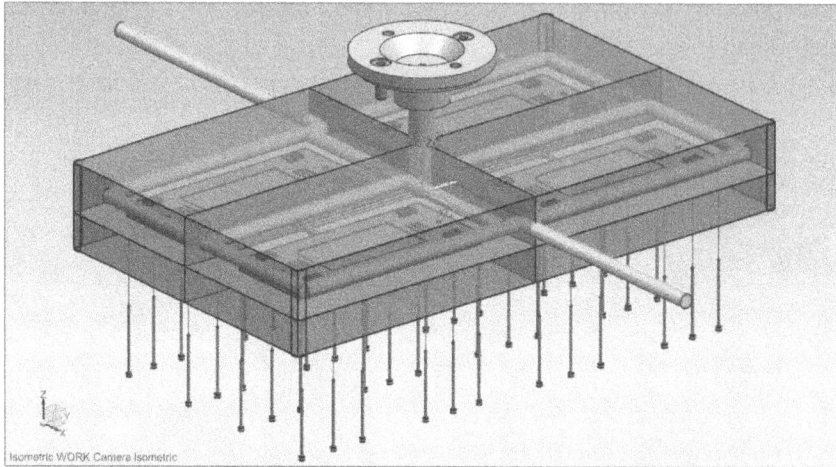

Figure 8-25 *The orientation and position of the cooling channel with core, cavity, runner, gate, sprue bush, and register ring*

4. Choose the **Extend Channel** tool from the **Cooling Tools** gallery of the **Mold Wizard** tab; the **Extend Channel** dialog box is displayed, refer to Figure 8-26. Also, you are prompted to select the cooling channel. Select the cooling channel; the **Distance** edit box appears on the screen. Enter **15** in the **Distance** edit box and then choose the **OK** button from the dialog box. Refer to Figure 8-27 for cooling channels after using the **Extend Channel** tool.

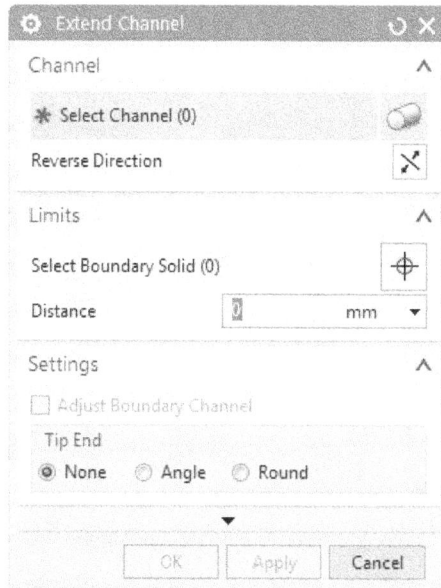

Figure 8-26 *The **Extend Channel** dialog box*

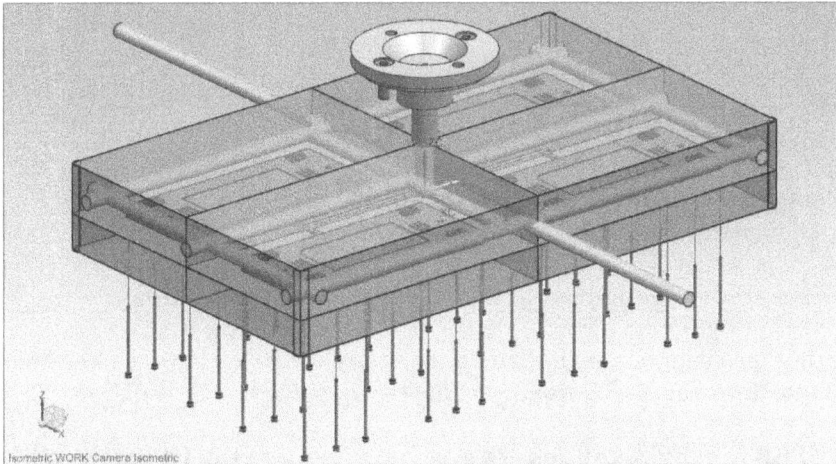

Figure 8-27 *The cooling channel after using the* ***Extend Channel*** *tool*

Creating the Pocket

Now, you will create pocket for the components to be placed or moved.

1. Choose the **Pocket** tool from the **Main** gallery of the **Mold Wizard** tab; the **Pocket** dialog box is displayed. Also, you are prompted to select the target bodies. Select the target bodies from the mold assembly.

2. Click in the **Select Object** area in the **Tool** rollout, and choose the tool body for which you need to create pocket. Next, choose the **Apply** button to create the pocket and the **Cancel** button to close the dialog box.

> **Note**
> *Tool body is the body which is used to subtract material from the given body. Target body is the body in which subtraction of material takes place.*

Saving and Closing the File

1. Choose **Menu > File > Close > Save and Close** from the **Top Border Bar** to save and close the file.

Tutorial 2

In this tutorial, you will add the ejector pin and cooling channel to the model (PCB_UPPER_CASE_top_###) contained in Tut 2 folder that you have downloaded. After adding the ejector pin and cooling channel, save the file.

(Expected time: 2 hr)

The following steps are required to complete this tutorial:

a. Start NX and open the file, refer to Figure 8-28.
b. Add ejector pin and cooling channel, refer to Figures 8-29 to 8-39.
c. Save the file.

Starting NX and Opening Mold Assembly

First, you need to start NX and then open a saved file.

1. Double-click on the NX shortcut icon on the desktop of your computer.

2. Choose the **Open** button from the **Standard** group of the **Home** tab or choose **Menu > File > Open** from the **Top Border Bar**; the **Open** dialog box is displayed.

3. Browse to **PCB_UPPER_CASE_top_###** in Tut 2 folder; the **PCB_UPPER_CASE_top_###** is displayed in the **File name** drop-down list. Then choose the **OK** button; the assembly is displayed, refer to Figure 8-28.

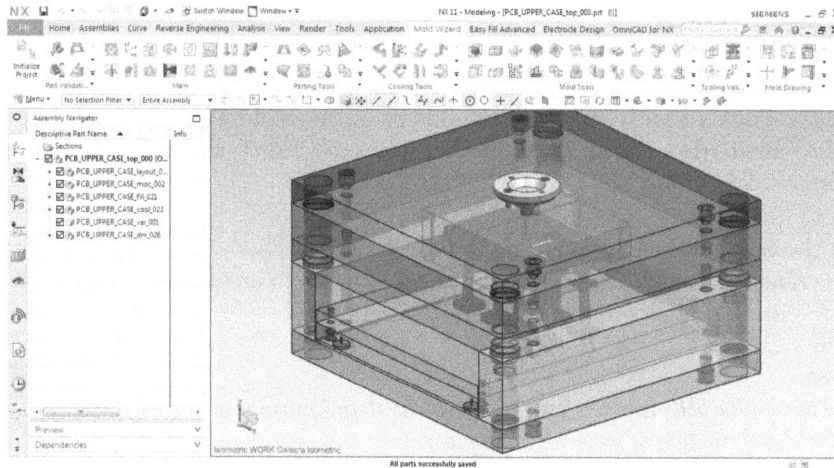

Figure 8-28 Mold assembly in Mold environment

Adding Ejector Pin to the Mold Assembly

Now, you will add the ejector pin to the mold assembly.

1. Choose **MW Standard Part Library** from the **Reuse Library**. Expand **MW Standard Part Library** and then choose the type of ejector pin from the **Ejection** folder of the **DME_MM** standard.

2. Double-click on **Ejector Pin[Shouldered]** in the **Member Select** panel; the **Standard Part Management** dialog box is displayed with the **Information** window, refer to Figure 8-29 (a) and 8-29 (b).

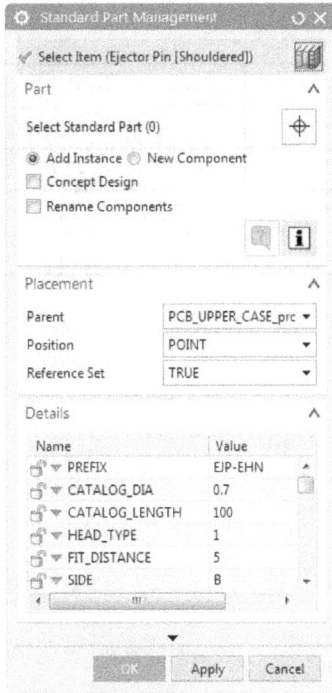

Figure 8-29 (a) The Standard Part Management dialog box

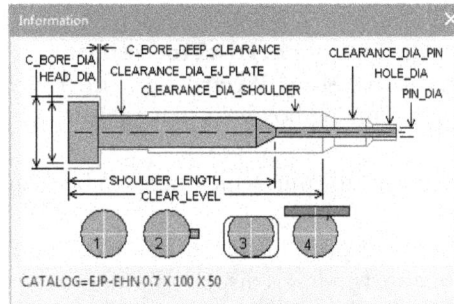

Figure 8-29 (b) The Information window

3. Select **EJP-EHN**, **0.7**, **125**, and **1** from the **PREFIX**, **CATALOG_DIA**, **CATALOG_LENGTH**, and **HEAD_TYPE** drop-down lists, respectively in the **Details** rollout. Choose the **Apply** button; the **Point** dialog box is displayed, refer to Figure 8-30. Specify the position of the ejector pin by placing the point. Choose the **Cancel** button to close the dialog box. Next, choose the **OK** button to close the **Standard Part Management** dialog box and the **Information** window. Refer to Figure 8-31 for the arrangement of ejector pin. Refer to Table 8-2 for the ejector pin positions.

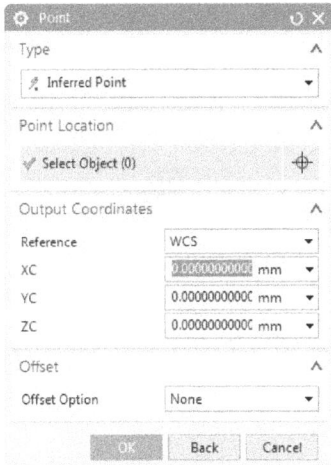

Figure 8-30 The Point dialog box

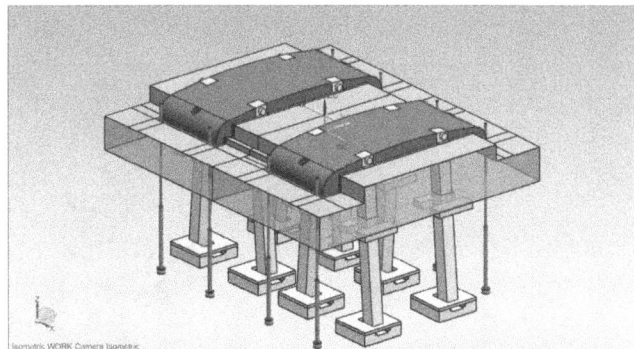

Figure 8-31 The arrangement of ejector pin in the mold assembly

Table 8-2 *Positions of ejector pin*

Ejector Pin	XC	YC
1	-85	77
2	-35	-77
3	-85	77
4	-35	77

Trimming the Ejector Pin

Now, you will trim the ejector pin in the mold assembly.

1. Choose the **Ejector Pin Post Processing** tool from the **Main** gallery of the **Mold Wizard** tab; the **Ejector Pin Post Processing** dialog box is displayed, refer to Figure 8-32.

2. Select the ejector pins which need to be trimmed.

3. Choose the **Apply** button to trim the ejector pin; the **Information** window is displayed. Choose the **Close** button to close the window. Choose the **Cancel** button to close the dialog box. Refer to Figure 8-33 for ejector pin after trimming.

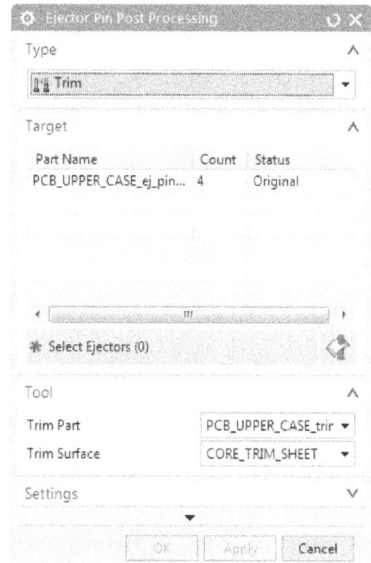

Adding the Cooling Channel

Now, you will add cooling channel for mold component.

1. Choose the **Pattern Channel** tool from the **Cooling Tools** gallery of the **Mold Wizard** tab; the **Pattern Channel** dialog box is displayed, refer to Figure 8-34.

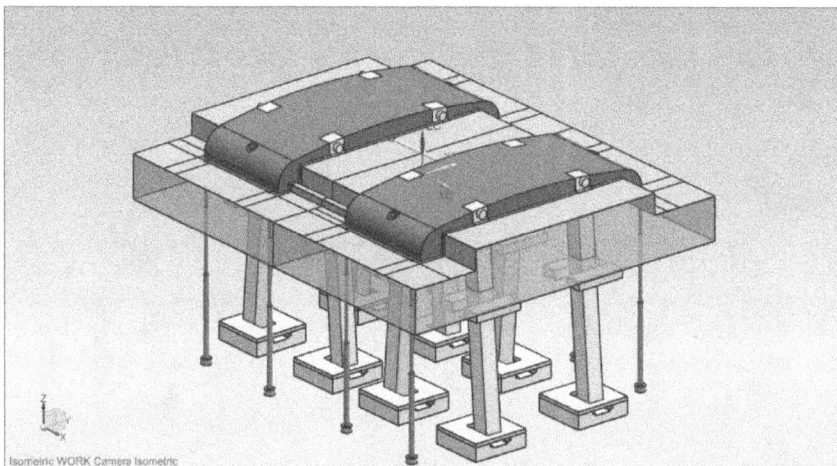

Figure 8-32 *The **Ejector Pin Post Processing** dialog box*

Figure 8-33 *The arrangement of ejector pin after trimming process in the mold assembly*

2. Choose the **Sketch Section** button in the **Channel Path** rollout; the **Create Sketch** dialog box is displayed, refer to Figure 8-35.

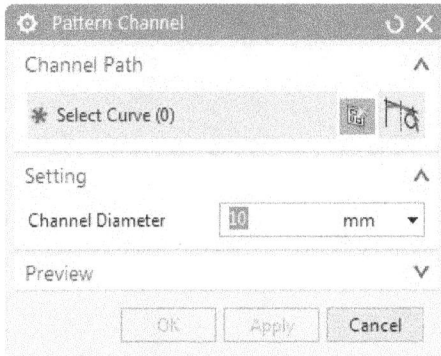

Figure 8-34 *The **Pattern Channel** dialog box*

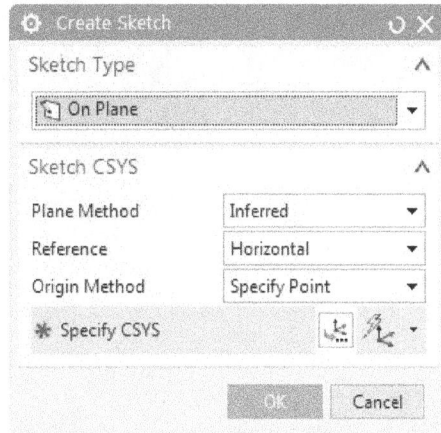

Figure 8-35 *The **Create Sketch** dialog box*

3. Create sketch for the cooling channel on a plane and then choose the **Finish** button to exit the sketch environment; the **Pattern Channel** dialog box is displayed. Enter **10** as the diameter of channel in the **Channel Diameter** edit box of the **Setting** rollout. Choose the **OK** button to close the dialog box. Refer to Figures 8-36 and 8-37 for understanding the position and orientation of the cooling channel.

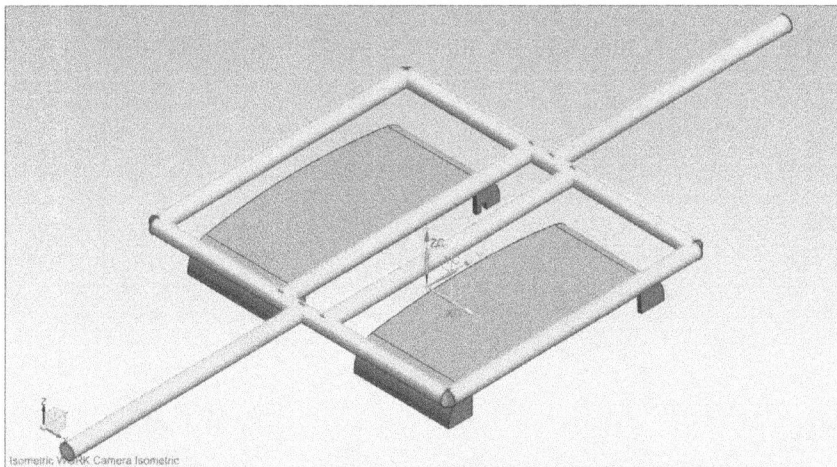

Figure 8-36 *The orientation and position of cooling channel*

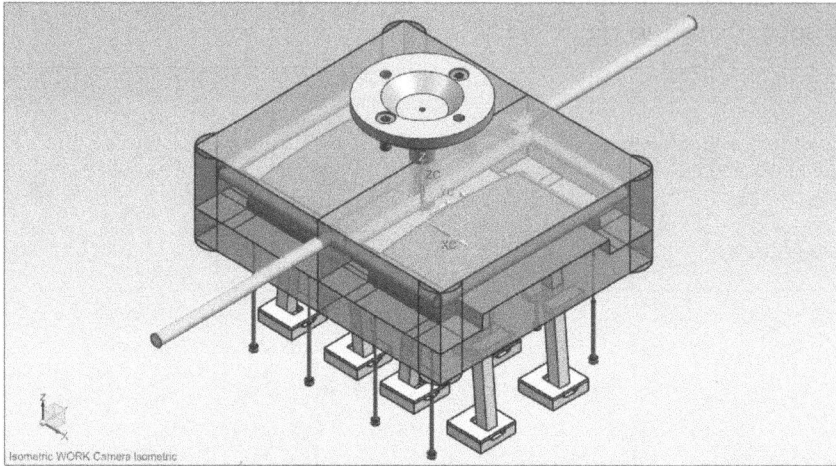

Figure 8-37 *The orientation and position of cooling channel with core, cavity, runner, gate, lifter, sprue bush, and register ring*

4. Choose the **Extend Channel** tool from the **Cooling Tools** gallery of the **Mold Wizard** tab; the **Extend Channel** dialog box is displayed, refer to Figure 8-38. Also, you are prompted to select the cooling channel. Select the cooling channel; the **Distance** edit box will be available on the screen. Enter **15** in the **Distance** edit box and then choose the **OK** button from the dialog box. Refer to Figure 8-39 for cooling channel after using the **Extend Channel** tool.

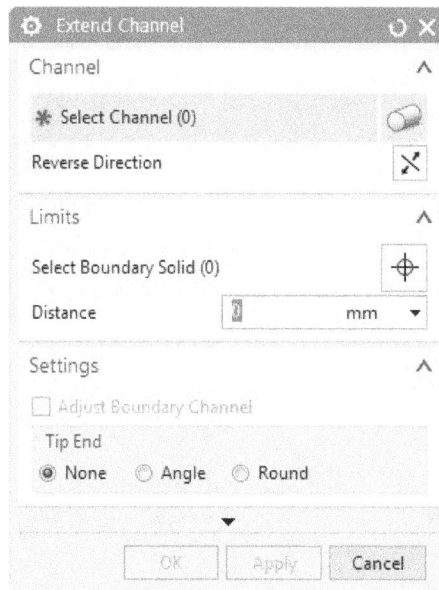

Figure 8-38 *The **Extend Channel** dialog box*

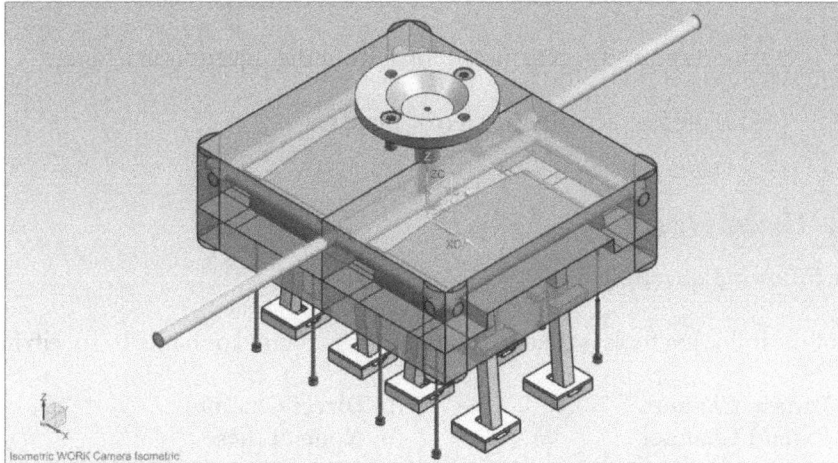

Figure 8-39 *The cooling channel after using the* **Extend Channel** *tool*

Creating the Pocket

Now, you will create pocket for the components to be placed or moved.

1. Choose the **Pocket** tool from the **Main** gallery of the **Mold Wizard** tab; the **Pocket** dialog box is displayed. Also, you are prompted to select the target bodies. Select the target bodies from the mold assembly.

2. Click in the **Select Object** area in the **Tool** rollout and choose the tool body for which you need to create pocket. Next, choose the **Apply** button to create pocket and the **Cancel** button to close the dialog box.

Saving and Closing the File

1. Choose **Menu > File > Close > Save and Close** from the **Top Border Bar** to save and close the file.

Self-Evaluation Test

Answer the following questions and then compare them to those given at the end of this chapter:

1. Which of the following tools is used to trim the ejector pin?

    (a) **Pocket**   (b) **Design Trim**
    (c) **Ejector Pin Post Processing**   (d) None of these

2. Which of the following tools is used to create the cooling channels using sketches or curves?

    (a) **Pattern Channel**   (b) **Direct Channel**
    (c) **Extend Channel**   (d) None of these

3. Ejector plate assembly is an assembly of _____, _____, and _____.

4. _____ plate receives direct actuated force from the injection machine.

5. _____ rod is used to actuate the ejector plate.

Review Questions

Answer the following questions:

1. Which of the following tools is used to create cooling channel or baffle by specifying a point?

    (a) **Pattern Channel**   (b) **Direct Channel**
    (c) **Extend Channel**   (d) None of these

2. Which of the following tools is used to adjust the position of the cooling channel?

    (a) **Extend Channel**   (b) **Pattern Channel**
    (c) **Adjust Channel**   (d) None of these

3. Which of the following tools is used to extend the length of the cooling channel?

    (a) **Extend Channel**   (b) **Pattern Channel**
    (c) **Adjust Channel**   (d) None of these

4. _____ is a process required to cool the component.

5. _____ plate is used to retain the ejector pin.

EXERCISES

Exercise 1

In this exercise, you will open the outout file (exr_01_top_###) of Exercise 1 of chapter 7. Add
ejector pin and cooling channel to the mold for the model shown in Figure 8-40 and save it.

(Expected time: 60 min)

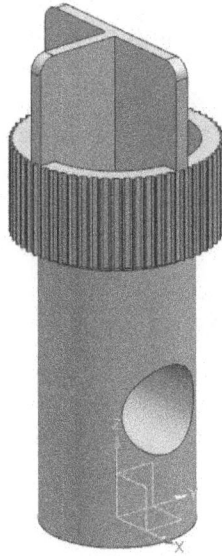

Figure 8-40 *Model for Exercise 1*

Exercise 2

In this exercise, you will open the outout file (exr_02_top_###) of Exercise 2 of chapter 7. Add ejector pin and cooling channel to the mold for the model shown in Figure 8-41 and save it.

(Expected time: 60 min)

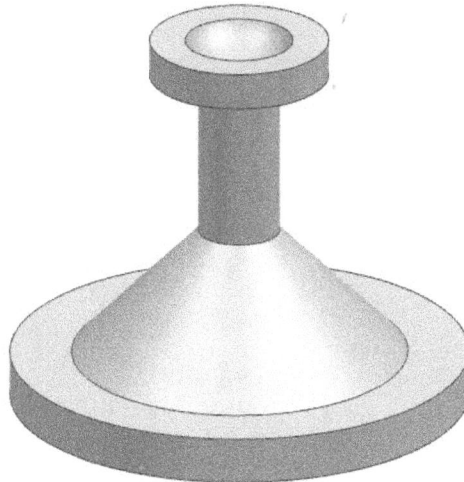

Figure 8-41 *Model for Exercise 2*

Answers to Self-Evaluation Test

1. c, **2.** a, **3.** ejector plate, retaining plate, and ejector rod, **4.** Ejector, **5.** Ejector

Chapter 9

Creating Electrodes

Learning Objectives

After completing this chapter, you will be able to:
- *Understand the Electrode Design environment*
- *Initialize electrode project*
- *Understand manufacturing geometry*
- *Understand bounding body*
- *Understand design blank*
- *Understand electrode fixture*

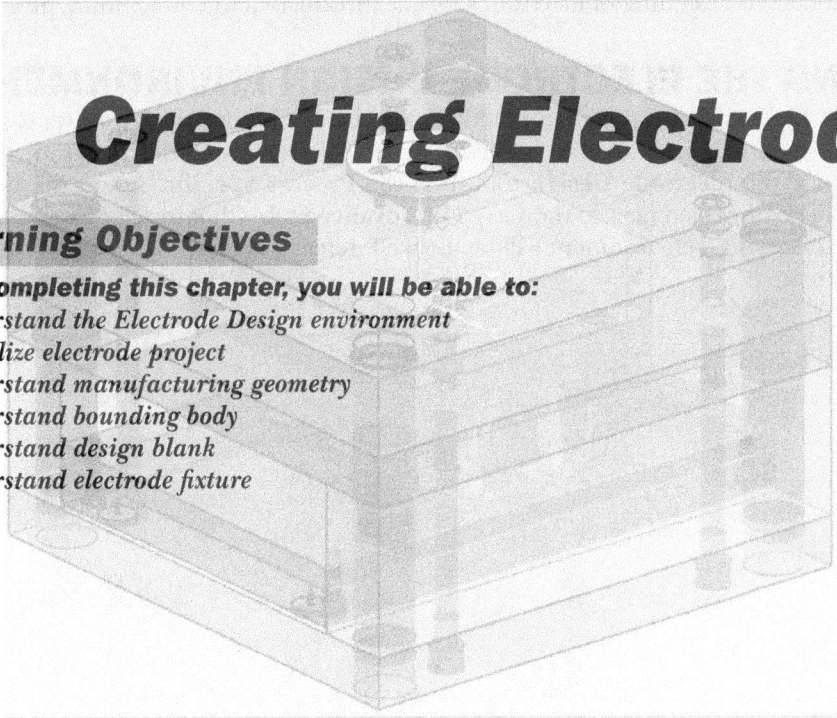

INTRODUCTION

In the previous chapter, you created ejection and cooling systems. In this chapter, you will learn how you can create electrodes using various tools.

ELECTRICAL DISCHARGE MACHINING

Electrical Discharge Machining is a process that uses electrodes to produce cavity features which are impossible to machine using milling tools. In NX, you can create electrodes by using tools available in Electrode Design application. This application provides a set of commands dedicated to quick and efficient designing of electrodes for Electrical Discharge Machining process.

INVOKING THE ELECTRODE DESIGN ENVIRONMENT

To invoke the Electrode Design environment, open the model on which you want to create electrodes in modeling environment, refer to Figure 9-1. Next, choose the **Application** tab and then choose the **Electrode Design** tool from the **Process Specific** group; the **Electrode Design** tab will be displayed next to the **Easy Fill Advanced** tab. Choose the **Electrode Design** tab; the electrode design environment will be invoked, refer to Figure 9-2. The tools in this tab are discussed next.

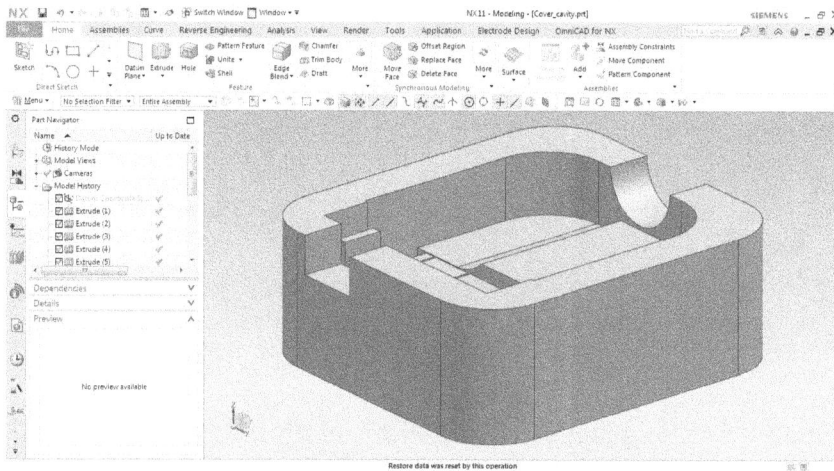

Figure 9-1 *The model opened in the Modeling environment*

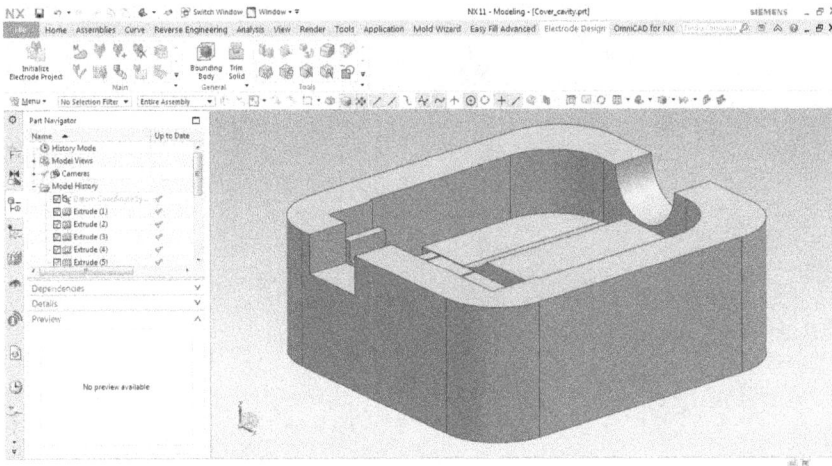

Figure 9-2 *The model opened in the **Electrode Design** environment*

Initialize Electrode Project

Ribbon: Electrode Design > Initialize Electrode Project

The **Initialize Electrode Project** tool helps to create electrode project while creating an electrode. To initialize an electrode project, choose the **Initialize Electrode Project** tool from the **Electrode Design** tab; the **Initialize Electrode Project** dialog box will be displayed, refer to Figure 9-3. Choose the **Add Machine Set** button from the **Machine Sets** rollout and choose the **OK** button; the machine set will be assigned for the electrode project and the project will be initialize.

Manufacturing Geometry

Ribbon: Electrode Design > Main > Manufacturing Geometry

The **Manufacturing Geometry** tool helps you to add manufacturing attributes to the faces of the model for downstream manufacturing. To do so, choose the **Manufacturing Geometry** tool from the **Main** gallery of the **Electrode Design** tab; the **Manufacturing Geometry** dialog box will be displayed, refer to Figure 9-4. Right-click on a machining method node such as EDM, WEDM, select the **New Group** option and then select faces; manufacturing attributes will be added to the faces of the model. Next, choose the **OK** button to close the dialog box.

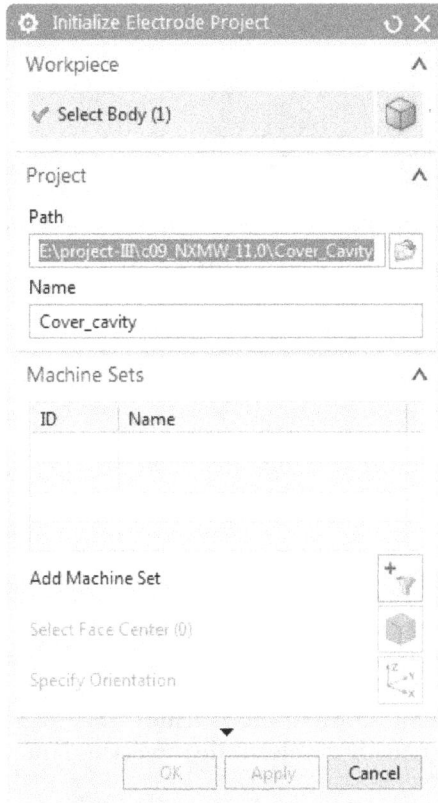

*Figure 9-3 The **Initialize Electrode Project** dialog box*

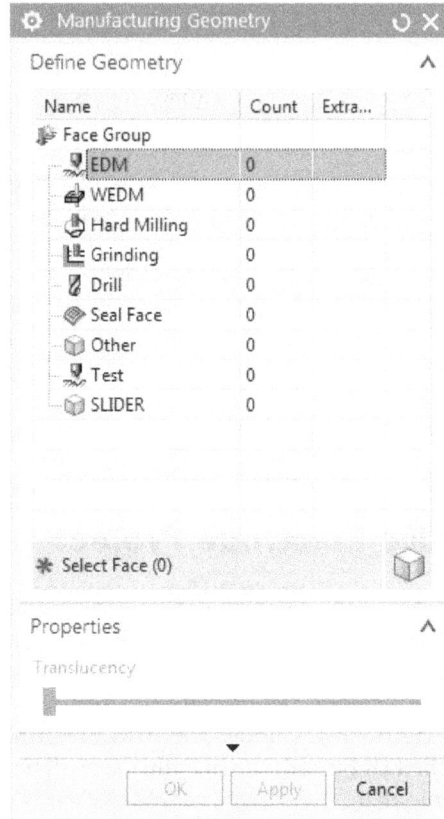

*Figure 9-4 The **Manufacturing Geometry** dialog box*

Bounding Body

Ribbon:	Electrode Design > General > Bounding Body

The **Bounding Body** tool helps you to create a box or a cylinder depending upon the option selected from the **Type** rollout. To create a bounding body, choose the **Bounding Body** tool from the **General** group of the **Electrode Design** tab; the **Bounding Body** dialog box will be displayed, refer to Figure 9-5. Select the faces and edit the size of the box by dragging the arrow handles; the bounding box will be created. Choose the **OK** button after editing the size.

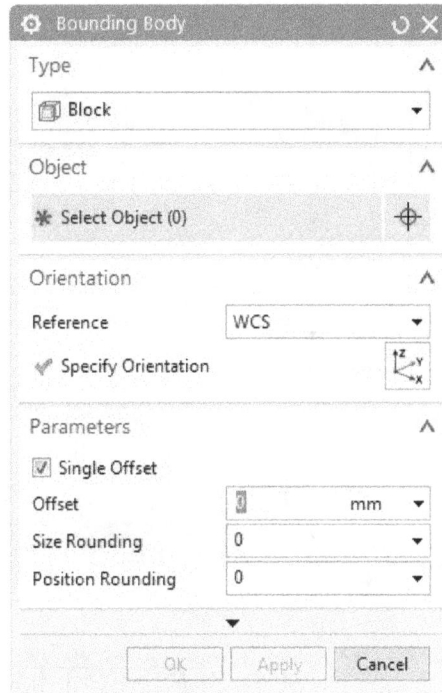

Figure 9-5 *The **Bounding Body** dialog box*

Note
*The **Bounding Body** tool helps you to create the electrode shape.*

Design Blank

Ribbon: Electrode Design > Main > Design Blank

The **Design Blank** tool helps you to add blank. To add a blank, choose the **Design Blank** tool from the **Main** gallery of the **Electrode Design** tab; the **Design Blank** dialog box will be displayed along with the **Information** window, refer to Figure 9-6. Select box or electrode; the blank will be created and then choose the **OK** button to close the dialog box.

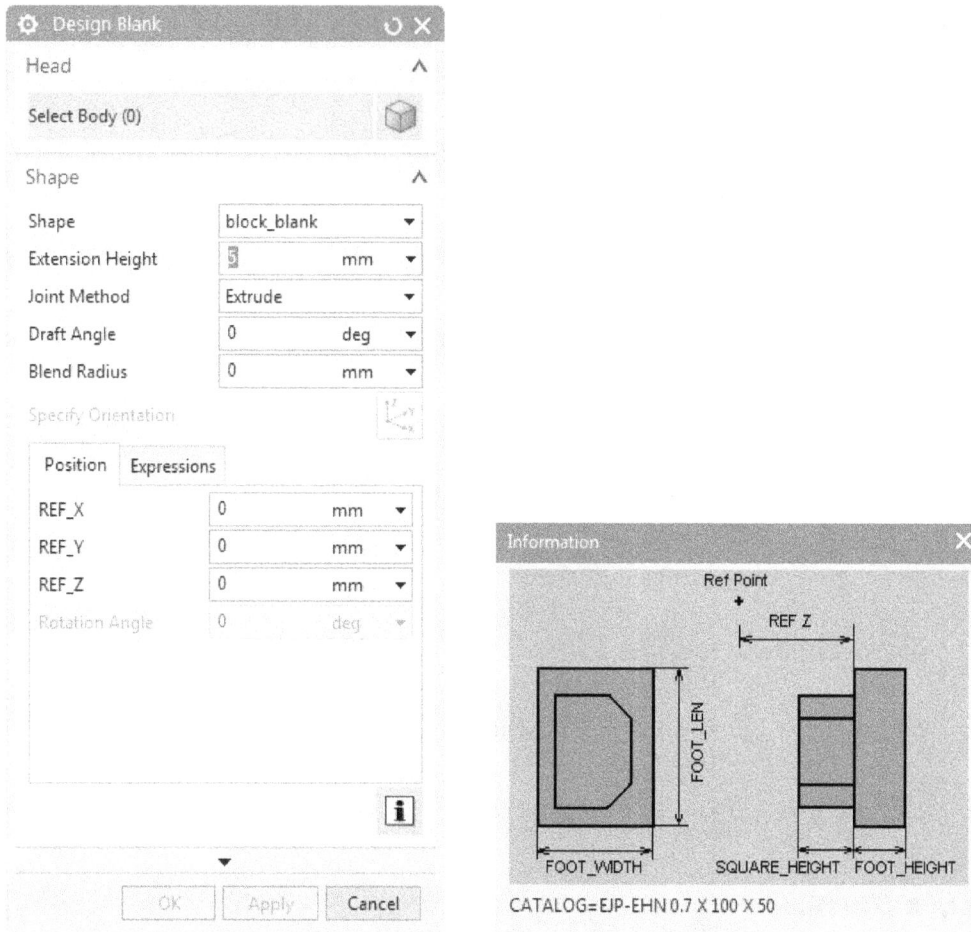

Figure 9-6 *The **Design Blank** dialog box along with the **Information** window*

Electrode Fixture

Ribbon:	Electrode Design > Main > Electrode Fixture

The **Electrode Fixture** tool helps you to add holder to blank. Choose the **Electrode Fixture** tool from the **Main** gallery of the **Electrode Design** tab; the **Electrode Fixture** dialog box along with the **Information** window will be displayed, refer to Figure 9-7. Choose the **Select Component** button from the **Placement** rollout and then select blank from the window; the holder will be added to blank. Choose the **OK** button to close the dialog box.

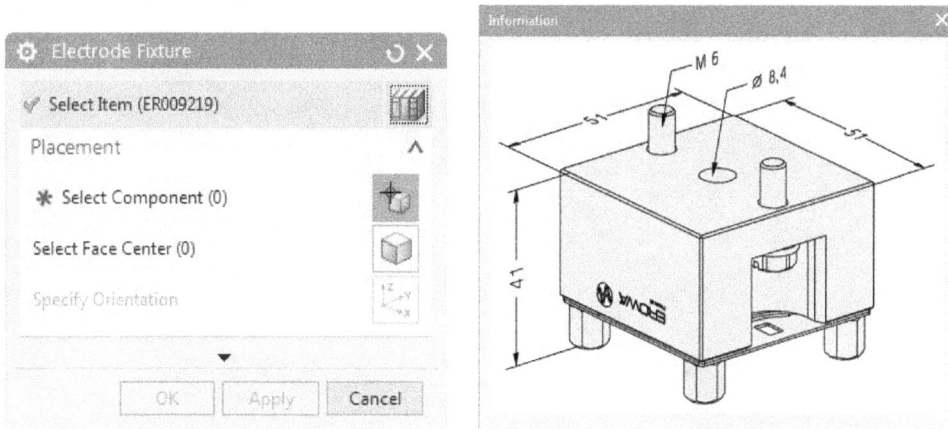

*Figure 9-7 The **Electrode Fixture** dialog box and the **Information** window*

Delete Body/Component

Ribbon: Electrode Design > Main > Delete Body/Component

The **Delete Body/Component** tool helps you to delete one or more selected electrode sparking bodies or electrode components. The components can be blanks, holders, or pallets. Electrode sparking bodies are also known as sparking heads. To delete electrode components, choose the **Delete Body/Component** tool from the **Main** gallery of the **Electrode Design** tab; the **Delete Body/Component** dialog box will be displayed, refer to Figure 9-8. By default, the **Sparking Body** option is selected in the **Type** drop-down list of the **Type** rollout. Select a sparking head from the window and then choose **Apply** and then the **Cancel** button.

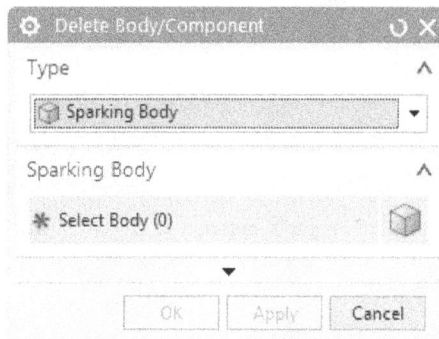

*Figure 9-8 The **Delete Body/Component** dialog box*

Check Electrode

Ribbon: Electrode Design > Main > Check Electrode

The **Check Electrode** tool helps you to check the interferences in the electrode assembly. If an electrode is out of alignment, it can gouge the core/cavity. Choose the **Check Electrode** tool from the **Main** gallery of the **Electrode Design** tab; the **Check Electrode** dialog box will be displayed, refer to Figure 9-9. Also, you will be prompted to select the workpiece to perform checking. Select the workpiece and then choose the **Select Electrode** button from the **Electrode** rollout and then choose the **OK** button; the **HD3D Tools** window will be displayed. You can check the result in the **Results** rollout.

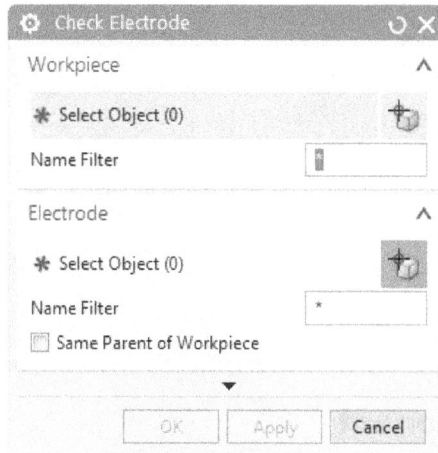

Figure 9-9 *The* ***Check Electrode*** *dialog box*

Electrode Bill of Material

Ribbon: Electrode Design > Main > Electrode Bill of Material

The **Electrode Bill of Material** tool helps you to manage the part attributes. To create a bill of material, choose the **Electrode Bill of Material** tool from the **Main** gallery of the **Electrode Design** tab; the **Bill of Material** dialog box will be displayed, refer to Figure 9-10. Also, you will be prompted to select the workpiece. Select the workpiece and then choose the green tick mark button. Next, choose the **Apply** button and then the **Export to Spreadsheet** button; the **Worksheet in BOM** window will be displayed which possess part information such as size, material. Save the file and choose the close button; the **Table Field** message box will be displayed. Choose the **OK** button and then again choose the **OK** button to close the dialog box.

The Bill of Material is commonly referred by acronym BOM. A bill of materials is automatically generated from the content with in the electrode assembly.

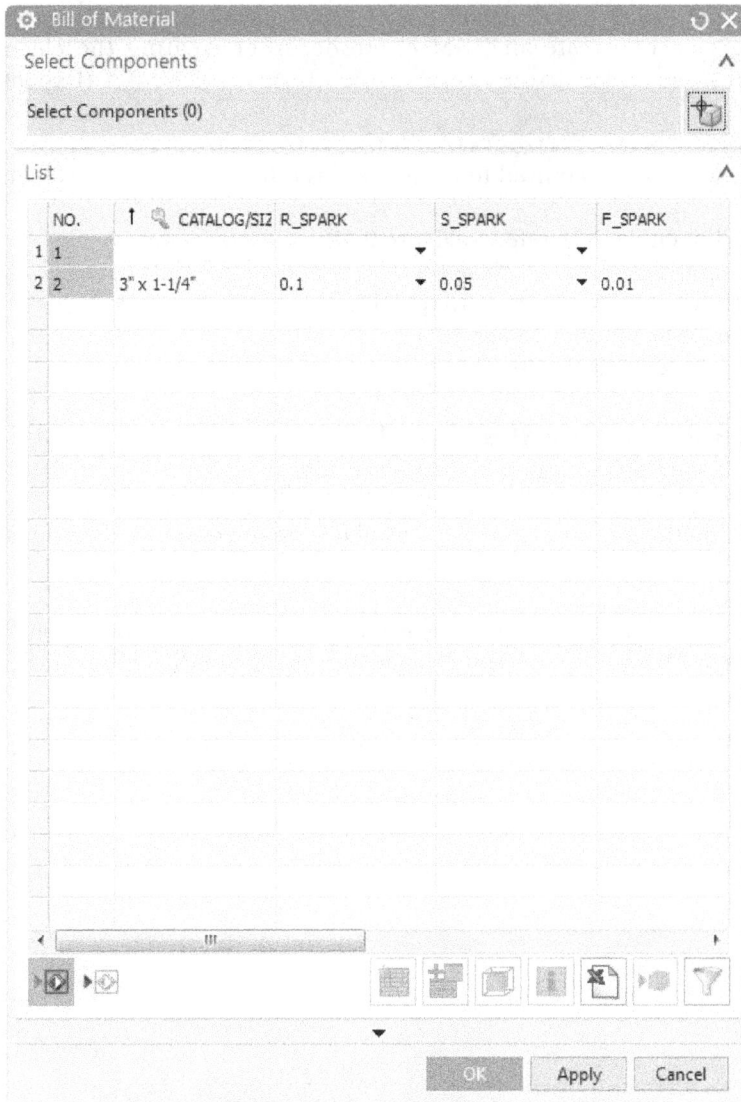

*Figure 9-10 The **Bill of Material** dialog box*

TUTORIAL

In these tutorials, you will create electrodes for the component. To perform the tutorials, you need to download the zipped file named as *c09_NX_Mold_input* from the **Input Files** section of the CADCIM website. The complete path for downloading the file is:

> *Textbooks > CAD/CAM > NX_Mold > Mold Design using NX 11.0: A Tutorial Approach > Input Files*

After the file is downloaded, extract the folder.

Tutorial 1

In this tutorial, you will create an EDM electrode. After creating the electrode, save the electrode design.   **(Expected time: 2 hr)**

The following steps are required to complete this tutorial:

a.   Start NX and open the file, refer to Figure 9-12.
b.   Initialize the project.
c.   Create electrode, refer to Figure 9-13 to 9-15.
d.   Create electrode head, refer to Figure 9-16.
e.   Create holder, refer to Figure 9-17.

Starting NX and Opening the Model

First, you need to start NX and then open a saved file.

1.   Double-click on the shortcut icon of NX available on the desktop of your computer to start NX.

2.   Choose the **Open** button from the **Standard** group of the **Home** tab or choose **Menu > File > Open** from the **Top Border Bar**; the **Open** dialog box is displayed refer to Figure 9-11.

Figure 9-11 *The **Open** dialog box*

3.   Browse to **Cover_cavity**; the **Cover_cavity** is displayed in the **File name** drop-down list. Then choose the **OK** button; the model is displayed, refer to Figure 9-12.

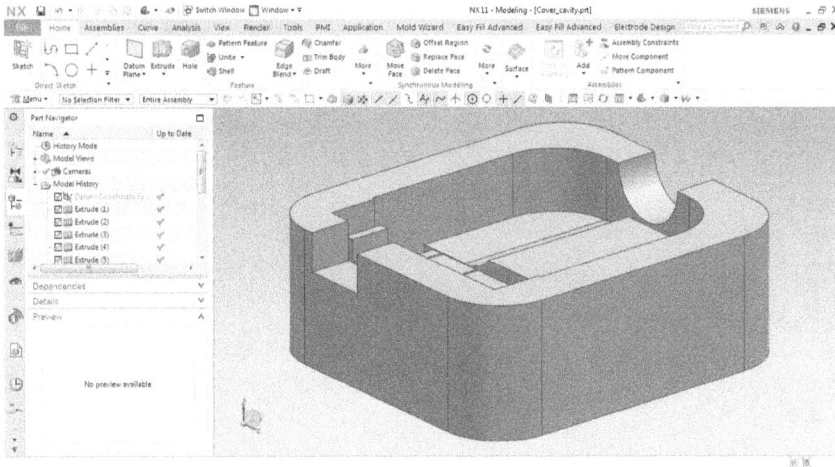

Figure 9-12 *The Cover_cavity model in Modeling environment*

Initializing the Project

Now, you will initialize the project.

1. Choose the **Initialize Electrode Project** tool from the **Electrode Design** tab; the **Initialize Electrode Project** dialog box is displayed.

2. Click on **Add Machine Set** in the **Machine Sets** rollout and choose the **OK** button.

Note
*You need to invoke the **Electrode Design** tab by clicking on the **Electrode Design** option in the Process Specific group from the **Application** tab.*

Manufacturing Geometry

Now, you will add manufacturing attributes for downstream manufacturing.

1. Double-click **Cover_cavity_working_###** from the **Assembly Navigator**. Choose the **Manufacturing Geometry** tool from the **Main Gallery** of the **Electrode Design** tab; the **Manufacturing Geometry** dialog box is displayed, refer to Figure 9-13. Right-click on **EDM** in the **Define Geometry** rollout and select the **New Group** option; the **electrode01** node is created, refer to Figure 9-14.

2. Select the faces of the model, refer to Figure 9-15. Choose the **OK** button to close the dialog box.

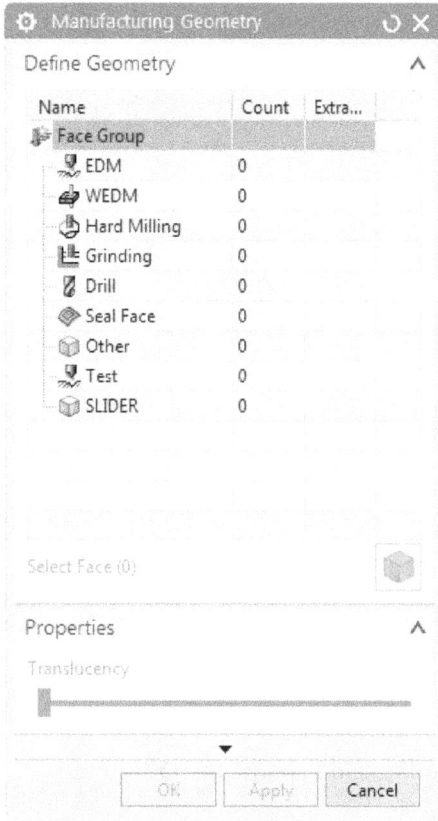

Figure 9-13 The **Manufacturing Geometry** dialog box

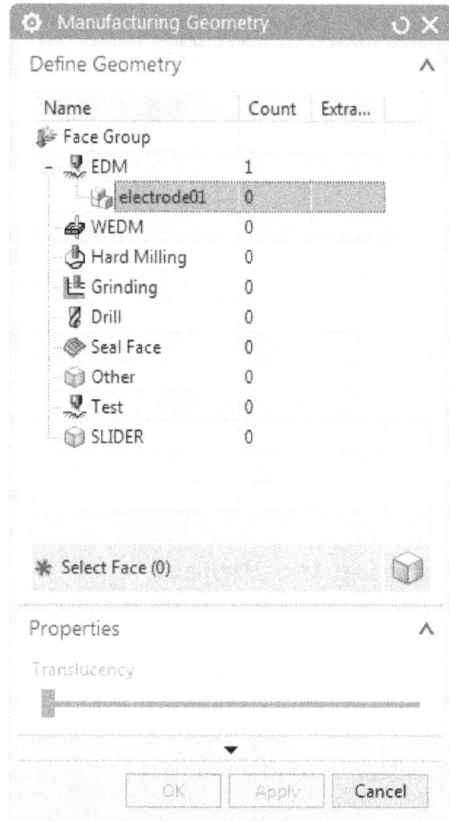

Figure 9-14 The **electrode 01** node in EDM

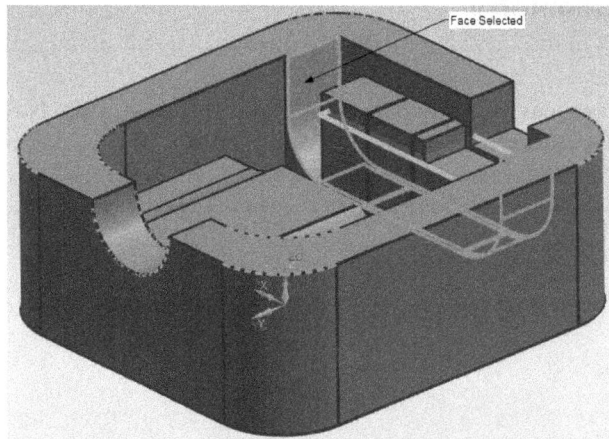

Figure 9-15 The faces selected in the model

Creating Electrode

1. Select **Cover_cavity_working_###** from the **Assembly Navigator**. Choose the **Bounding Body** tool from the **General** group of the **Electrode Design** tab; the **Bounding Body** dialog box is displayed. Select the faces of the model, refer to Figure 9-16(a) and 9-16(b). Adjust the size of the electrode by dragging the corresponding arrow handle, refer to Figure 9-17.

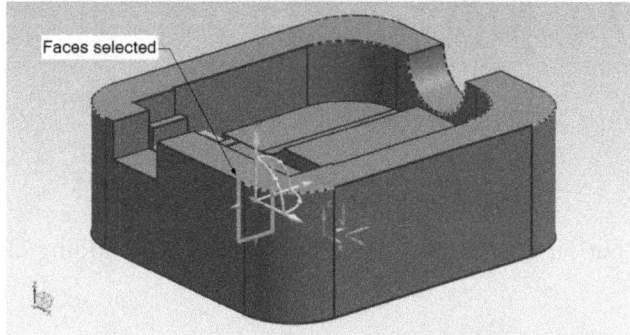

Figure 9-16(a) Face selected for electrode

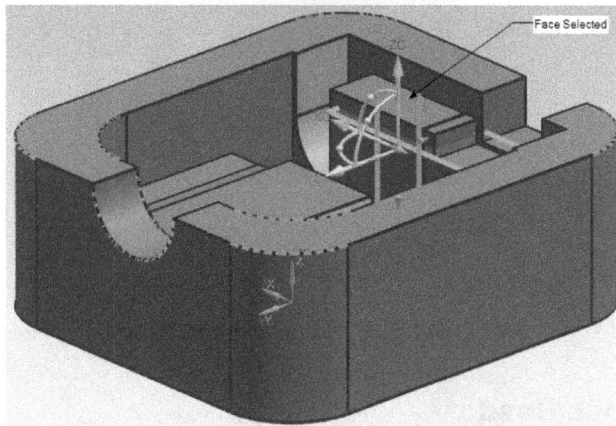

Figure 9-16(b) Face selected for electrode

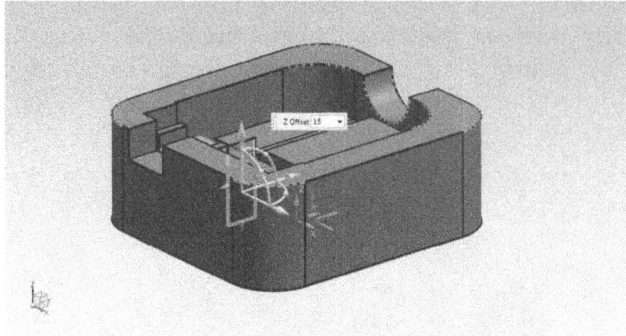

Figure 9-17 *The electrode after resizing*

2. Choose the **OK** button to close the dialog box. Model after adding electrode is shown in Figure 9-18.

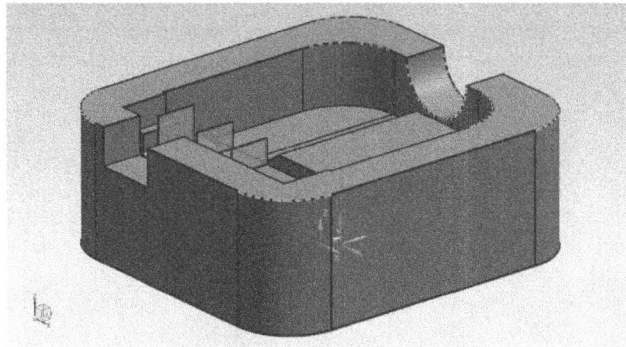

Figure 9-18 *The electrode arrangement*

Creating Electrode Head

1. Choose the **Design Blank** tool from the **Main Gallery** of the **Electrode Design** tab; the **Design Blank** dialog box is displayed.

2. Select all the electrodes and choose the **OK** button. Refer to Figure 9-19 for electrode head attachment.

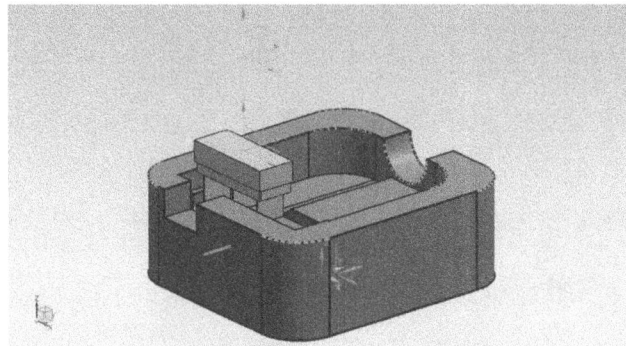

Figure 9-19 *The electrode head*

Creating Holder

1. Choose the **Electrode Fixture** tool from the **Main Gallery** of the **Electrode Design** tab; the **Electrode Fixture** dialog box is displayed.

2. Select the **Select Component** button from the **Placement** rollout and then select the electrode. Choose the **OK** button. Refer to Figure 9-20 for holder attachment.

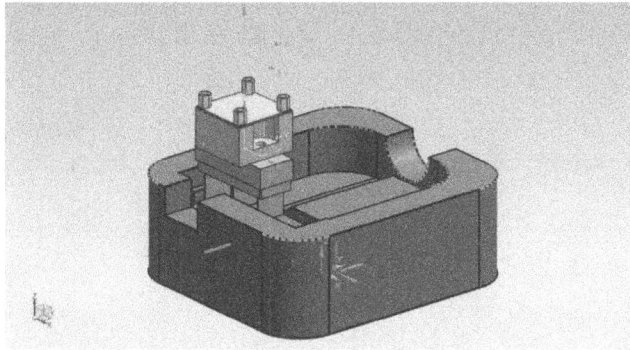

Figure 9-20 The Holder attachment

Note

*If you does not want to create a reference from the fixture template, clear the **Reference** check box from the **Settings** rollout of the **Electrode Fixture** dialog box .*

Saving and Closing the File

1. Choose **Menu > File > Save** from the **Top Border Bar**, the **Save CGM** message box is displayed. Choose the **Yes** button to close the message box.

Self-Evaluation Test

Answer the following questions and then compare them to those given at the end of this chapter:

1. Electrical Discharge Machining uses _____ to produce cavity feature.

2. The _____ tool is used to invoke the electrode design environment.

3. The _____ tool is used to create electrode project.

4. The _____ tool helps you to add manufacturing attributes to the face of the model for downstream manufacturing.

Review Questions

Answer the following questions:

1. The _____ tool is used to a add blank.

2. The _____ tool is used to a add fixture.

EXERCISE

To perform the exercise, you need to download the zipped file named as *c09_NX_Mold_input* from the **Input Files** section of the CADCIM website. The complete path for downloading the file is:

> *Textbooks > CAD/CAM > NX_Mold > Mold Design using NX 11.0: A Tutorial Approach > Input Files*

After the file is downloaded, extract the folder.

Exercise 1

In this exercise, you need to open the model that you have downloaded and then create electrode for the model shown in Figure 9-21.   **(Expected time: 60 min)**

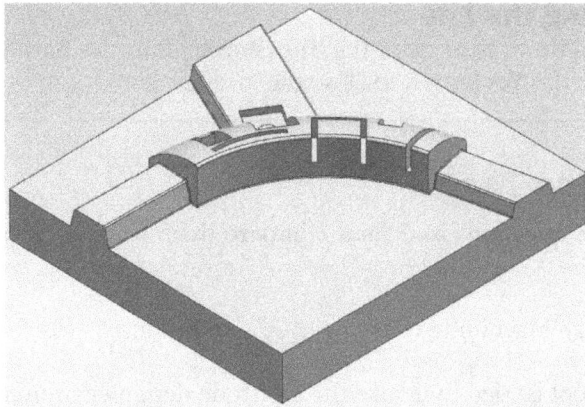

Figure 9-21   *The model for electrode designing*

Answers to Self-Evaluation Test
1. electrode, **2. Electrode Design, 3. Initialize Electrode Project, 4. Manufacturing Geometry**

Chapter 10

Documentation

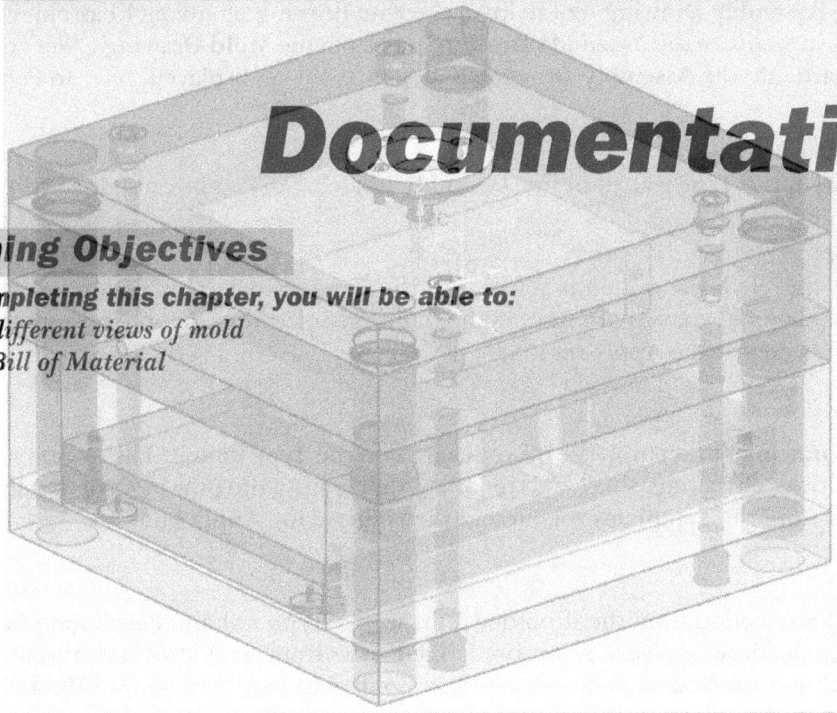

INTRODUCTION

In the previous chapters, you have learned to create mold. In this chapter, you will learn to create a detail drawing of the mold for manufacturing. Also, you will learn to use tools for creating detail drawings.

CREATING ASSEMBLY DRAWING

Ribbon: Mold Wizard > Mold Drawing gallery > Assembly Drawing

The **Assembly Drawing** tool helps you create drawing of any mold assembly. To create drawing, choose the **Assembly Drawing** tool from the **Mold Drawing** gallery of the **Mold Wizard** tab; the **Assembly Drawing** dialog box will be displayed, refer to Figure 10-1.

Type Rollout

The options in the **Type** rollout in the **Assembly Drawing** dialog box are discussed next.

Visibility

By default, the **Visibility** option is selected in the **Type** rollout. This option helps you select the parts of the mold you want to show in the drawing. Select the component from the drawing area and then choose the **Apply** button.

Drawing

Select the **Drawing** option from the drop-down list in the **Type** rollout. This option helps you to specify the type of drawing, sheet, and template for creating the drawing. Next, select the type of template from the **Templates** rollout and then choose the **Apply** button.

View

Select the **View** option from the drop-down list in the **Type** rollout. This option helps you to add views to the drawing. Next, select predefined views from the **View Control** rollout. Specify the scale value in the **Scale** edit box of the **View Control** rollout. Choose the **OK** button to close the dialog box.

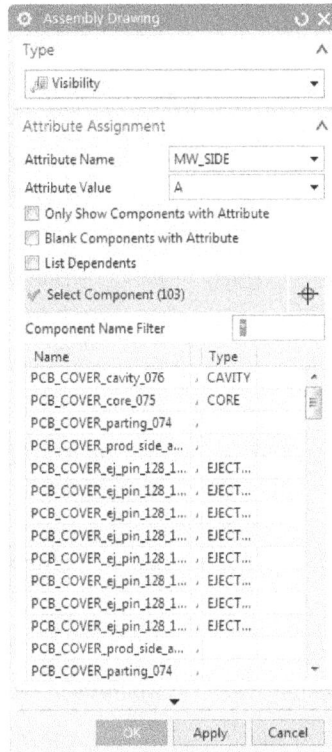

Figure 10-1 The **Assembly Drawing** dialog box

Component Drawing

Ribbon: Mold Wizard > Mold Drawing gallery > Component Drawing

The Component Drawing tool helps you to create and manage drawing components in a mold assembly. To create a drawing, choose the **Component Drawing** tool from the **Mold Drawing** gallery of the **Mold Wizard** tab; the **Component Drawing** dialog box will be displayed, refer to Figure 10-2. Select the components from the drawing window and then choose the **Create Drawing** button from the **Create Drawing** area of the **Drawing** rollout; the drawing will be generated. Also, the **New Iray + Ray Traced Studio Rendering** message box and the **Information** window will be displayed. Choose the **OK** and **Close** buttons to close the message box and window respectively. Then choose the **Close** button to close the dialog box.

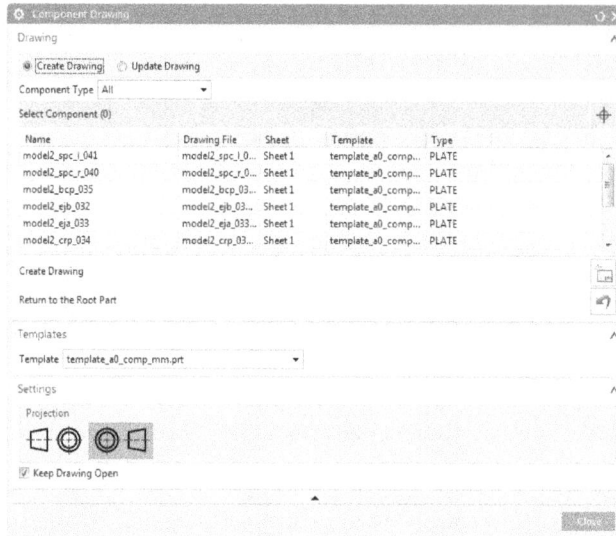

Figure 10-2 The **Component Drawing** *dialog box*

Bill of Material

Ribbon: Mold Wizard > Main gallery > Bill of Material

A bill of material is a list of components in which the specifications of the components is mentioned. To generate the bill of material, choose the **Bill of Material** tool from the **Main** gallery of the **Mold Wizard** tab; the **Bill of Material** dialog box will be displayed, refer to Figure 10-3. Select the components from the drawing window which you want to add to the bill of material list. Choose the **Export to Spreadsheet** button from the **Bill of Material** dialog box; an excel worksheet will be displayed, refer to Figure 10-4. You can also edit the bill of material list according to standard parts name. Choose the **Save** button to save the excel worksheet. Choose the **Close** button; the **Table Field** window will be displayed. Choose the **OK** button to close the window. Choose the **OK** button to close the **Bill of Material** dialog box.

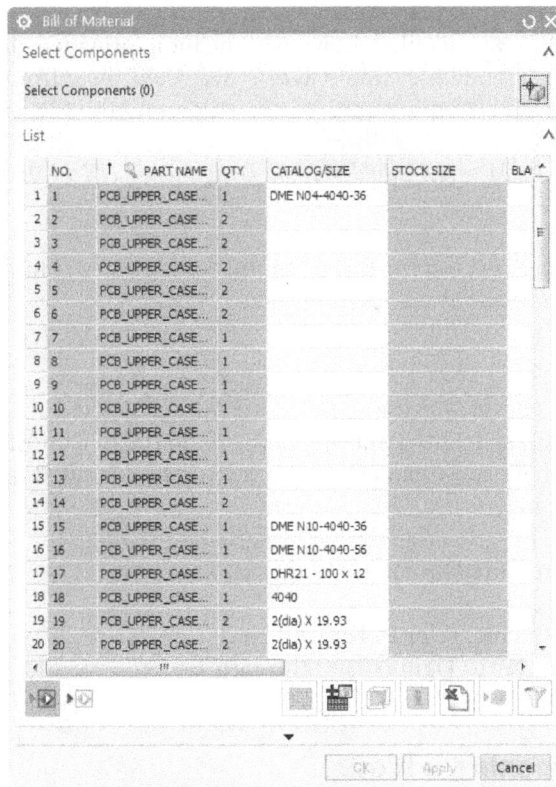

Figure 10-3 The **Bill of Material** *dialog box*

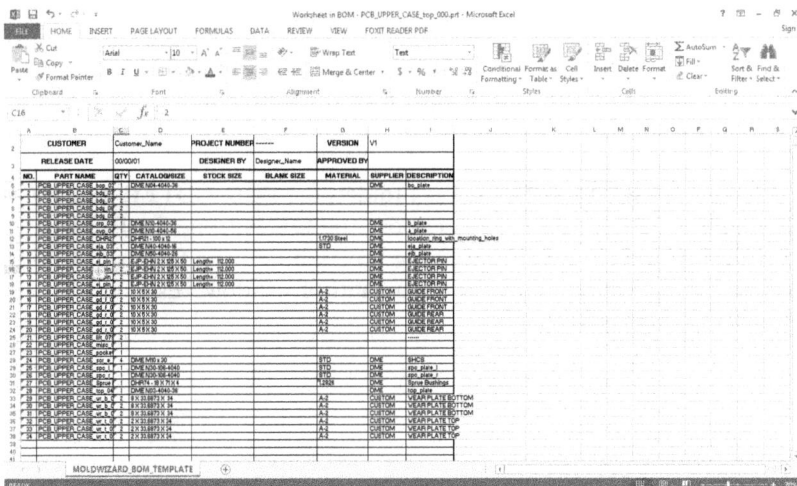

Figure 10-4 The **Bill of Material** *worksheet*

TUTORIALS

In these tutorials, you will create detailed drawing of the mold provided. To perform the tutorials, you need to download the zipped file named as *c10_NX_Mold_input* from the Input Files section of the CADCIM website. The complete path for downloading the file is:

> *Textbooks > CAD/CAM > NX_Mold > Mold Design using NX 11.0: A Tutorial Approach > Input Files*

After the file is downloaded, extract the folder. In this folder, you will find Tut1 and Tut2 folders containing input files for Tutorial 1 and Tutorial 2.

Tutorial 1

In this tutorial, you will create detailed drawing of the mold (PCB_COVER_top_###) contained in Tut 1 folder that you have downloaded. After creating the detailed drawing, save the file.

(Expected time: 2 hr)

The following steps are required to complete this tutorial:

a. Start NX and open the file, refer to Figure 10-6.
b. Generate the views, refer to Figures 10-9 through 10-12.
c. Save the file.

Starting NX and Opening a Model

First, you need to start NX and then open a saved file.

1. Double-click on the shortcut icon of NX available on the desktop of your computer to start NX.

2. Choose the **Open** button from the **Standard** group of the **Home** tab or choose **Menu > File > Open** from the **Top Border Bar**; the **Open** dialog box is displayed, refer to Figure 10-5.

3. Browse to **PCB_COVER_top_###** in Tut 1 folder; the **PCB_COVER_top_###** is displayed in the **File name** drop-down list. Next, choose the **OK** button; the **Information** window is displayed with the **Update event list** dialog box. Choose the **OK** and **X** buttons from the **Update event list** dialog box and the **Information** window, respectively; the assembly is displayed, refer to Figure 10-6.

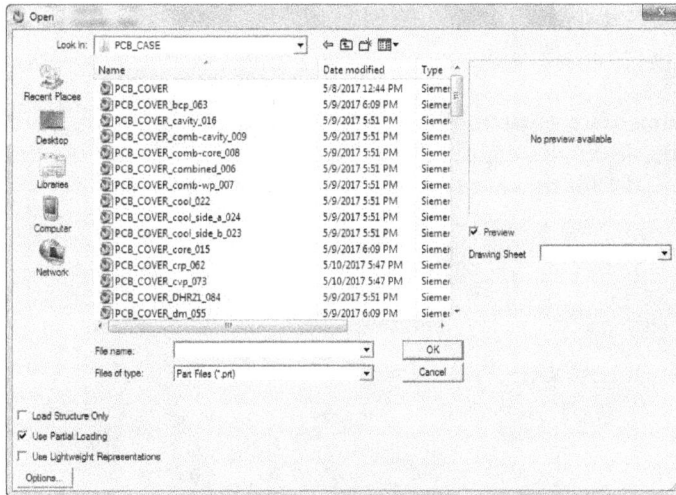

Figure 10-5 The **Open** *dialog box*

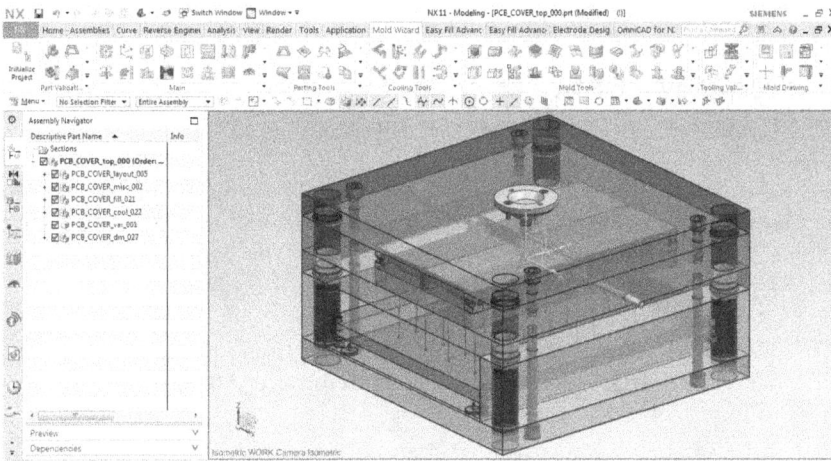

Figure 10-6 The **PCB_COVER** *assembly*

Hiding the Components of the Mold

Now, you need to hide the components of the mold.

1. Choose the **Immediate Hide** tool from the **Show Hide** gallery of the **Visibility** group of the **View** tab; the **Immediate Hide** dialog box is displayed. Select the fixed half for hiding. Refer to Figure 10-7 for moving half of the mold.

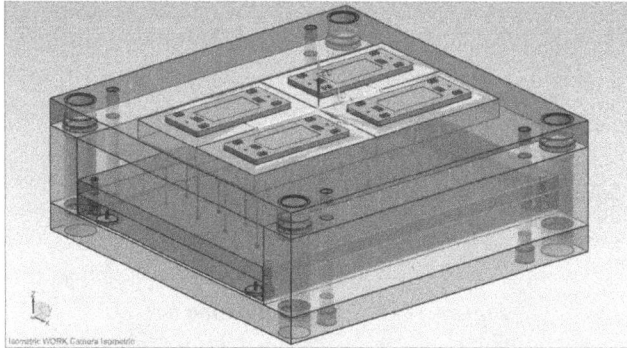

Figure 10-7 *The moving half of the mold*

Generating the Views of the Mold

Now, you need to generate the views of the mold.

1. Choose the **Assembly Drawing** tool from the **Mold Drawing** gallery of the **Mold Wizard** tab; the **Assembly Drawing** dialog box is displayed, refer to Figure 10-8. Select all the components from the drawing window and then choose the **Apply** button; the **Information** window is displayed. Choose the **Close** button to close the window.

2. Choose the **Drawing** option from the **Type** rollout and then select the **Self Contained** radio button from the **Drawing Type** rollout. Select **template_A0_asy_fam_mm.prt** from the **Templates** rollout.

3. Choose the **Apply** button and then the **Cancel** button to close the dialog box.

4. Choose the **Application** tab. Next, choose the **Drafting** tool from the **Design** group; the drafting environment is activated.

5. Choose the **Base View** tool from the **View** group; the **Base View** dialog box is displayed. Select **1:2** from the **Scale** drop-down list in the **Scale** rollout and then click to place the view. Next, add the required dimensions to the view, refer to Figure 10-9.

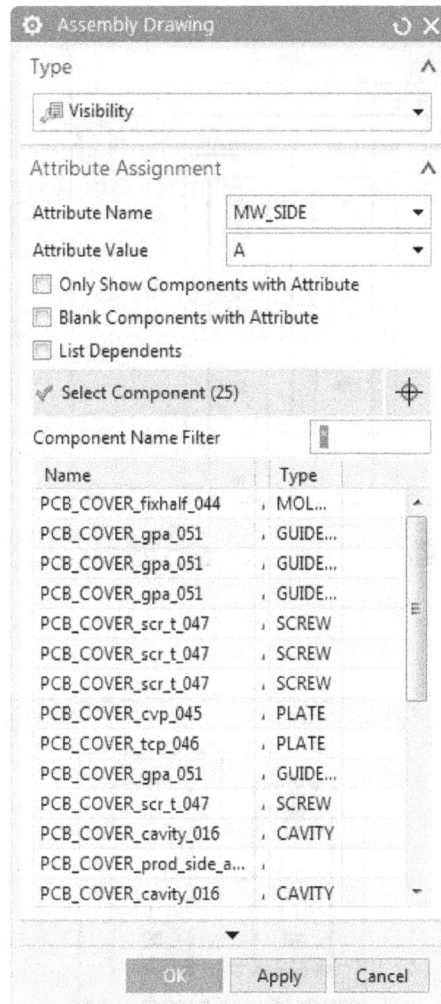

*Figure 10-8 The **Assembly Drawing** dialog box*

6. Create section line and section view for the moving half, as shown in Figures 10-10 and 10-11. Refer to NX11.0 for Designers by Prof. Sham Tickoo for details of section line and section view tools.

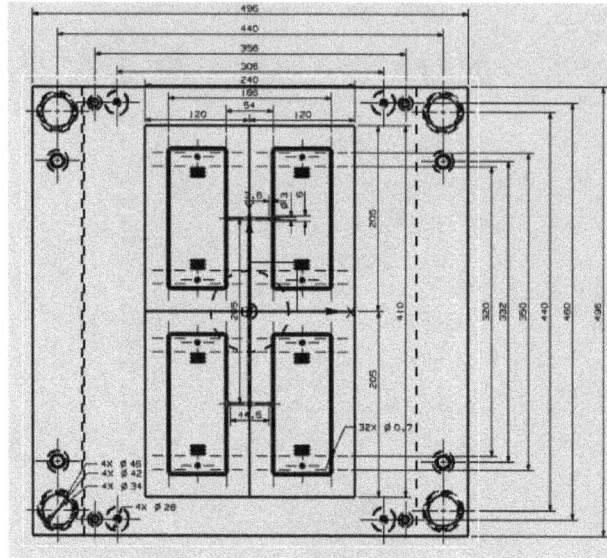

Figure 10-9 *The dimensioning of moving half*

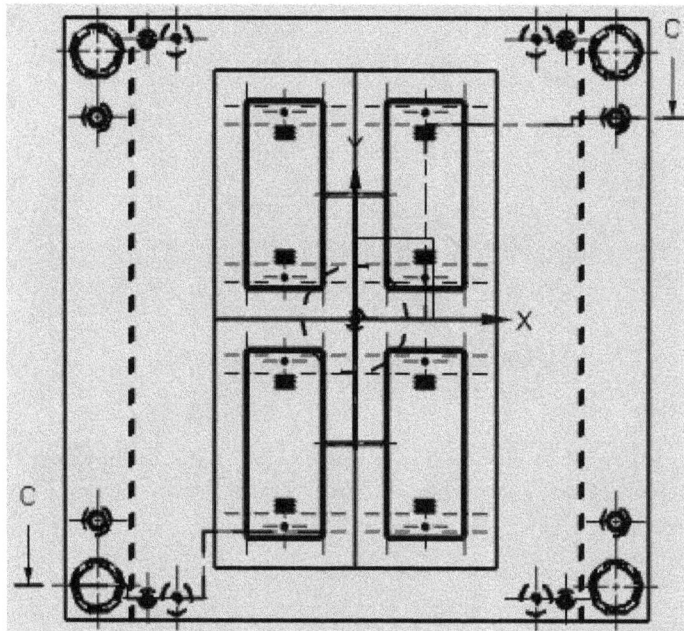

Figure 10-10 *Section line on the moving half*

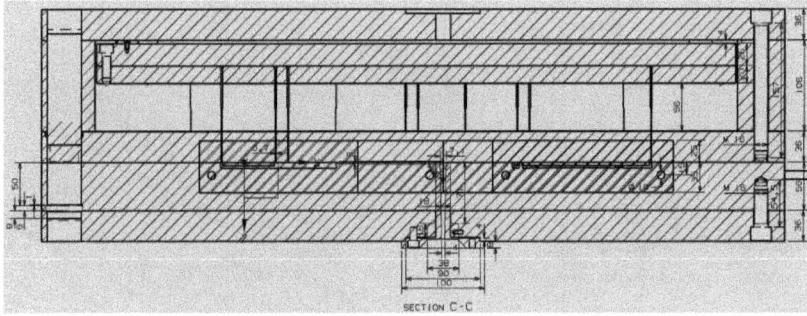

Figure 10-11 *Section view of the mold*

7. Repeat the same procedure for detailing the other views, refer to Figures 10-12 and 10-13.

Note

Before creating the section view, you need to unhide all the components of the mold in the mold wizard environment.

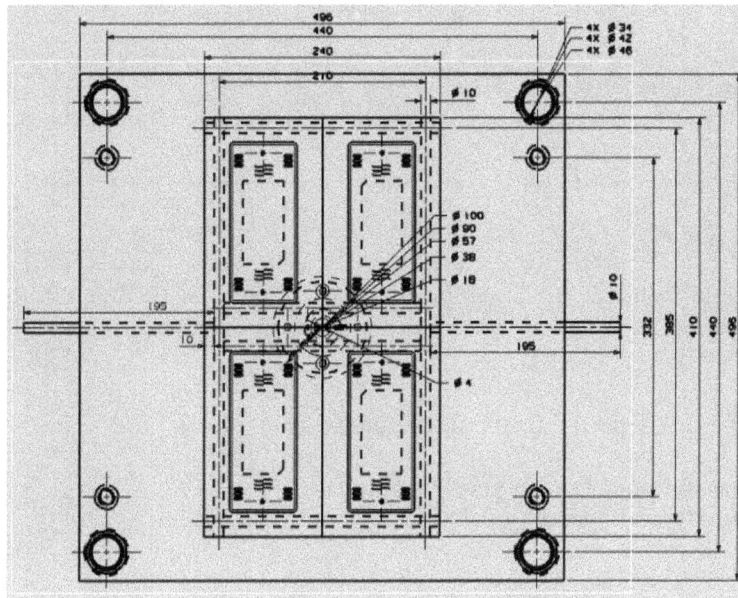

Figure 10-12 *Inverted top view of the stationary half*

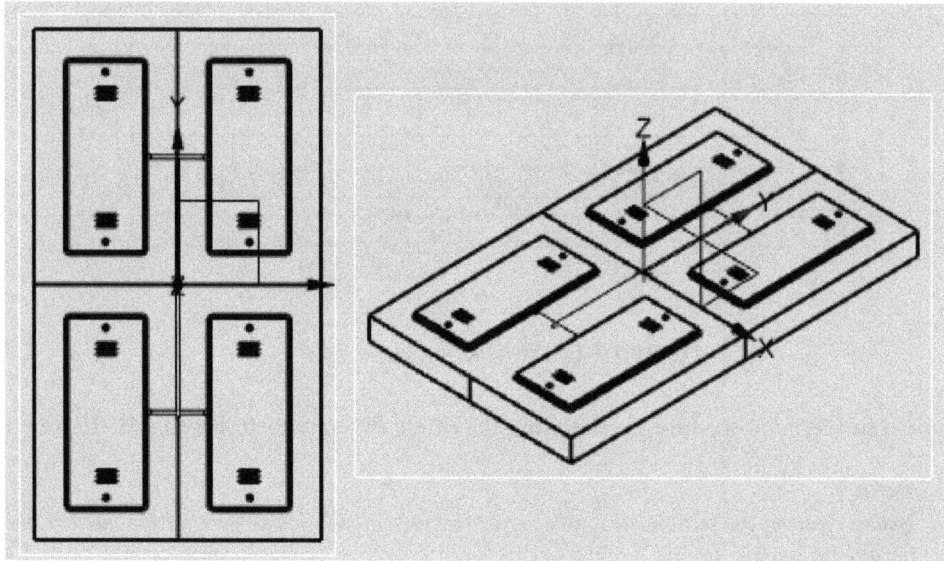

Figure 10-13 Top and isometric views of the core insert

Saving and Closing the File
1. Choose **Menu > File > Close > Save and Close** from the **Top Border Bar** to save and close the file.

Tutorial 2

In this tutorial, you will create detail drawing of the mold (PCB_UPPER_CASE_top_###) contained in Tut 2 folder that you have downloaded. After creating the detail drawing, save the file.   **(Expected time: 2 hr)**

The following steps are required to complete this tutorial:

a. Start NX and open the file, refer to Figure 10-14.
b. Generate the views, refer to Figures 10-18 through 10-23.
c. Save the file.

Starting NX and Opening a Model
First, you need to start NX and then open a saved file.

1. Double-click on the shortcut icon of NX available on the desktop of your computer to start NX.

2. Choose the **Open** button from the **Standard** group of the **Home** tab or choose **Menu > File > Open** from the **Top Border Bar**; the **Open** dialog box is displayed, refer to Figure 10-14.

Open

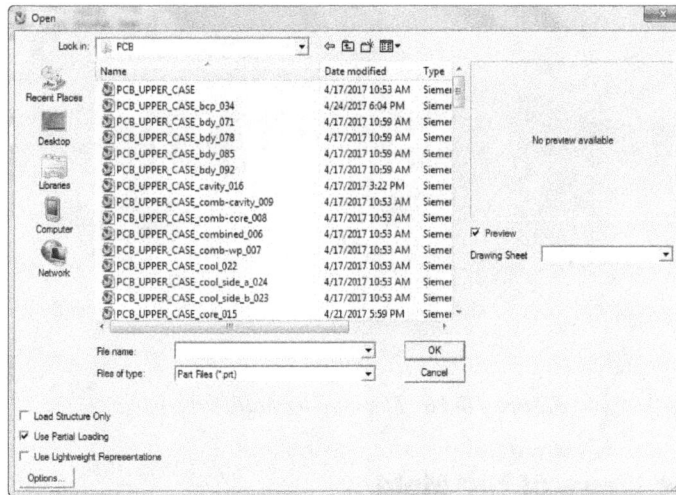

*Figure 10-14 The **Open** dialog box*

3. Browse to **PCB_UPPER_CASE_top_###** in Tut 2 folder; **PCB_UPPER_CASE_top_###**
    is displayed in the **File name** drop-down list. Next, choose the **OK** button; the **Information**
    window is displayed with the **Update event list** dialog box. Choose the **OK** and **X** buttons
    from the **Update event list** dialog box and the **Information** window, respectively; the assembly
    is displayed, refer to Figure 10-15.

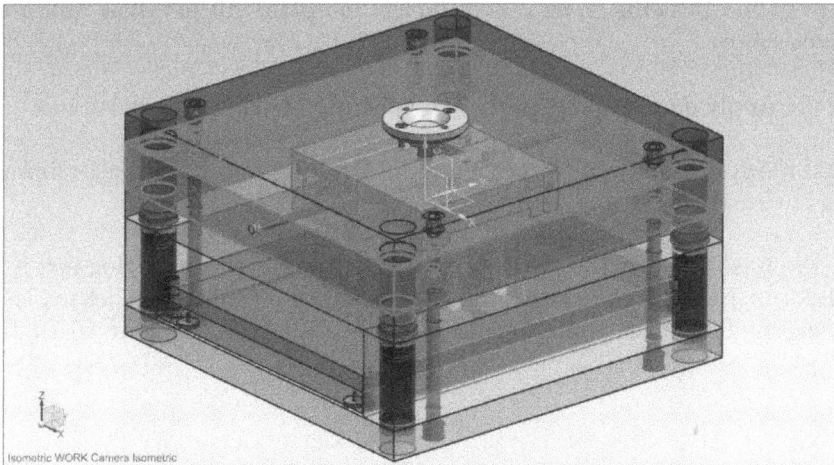

*Figure 10-15 The **PCB_UPPER_CASE** mold design*

Hiding the Components of the Mold

Now, you need to hide the components of the mold.

1. Choose the **Immediate Hide** tool from the **Show Hide** gallery of the **Visibility** group of
    the **View** tab; the **Immediate Hide** dialog box is displayed. Select the fixed half for hiding,
    refer to Figure 10-16.

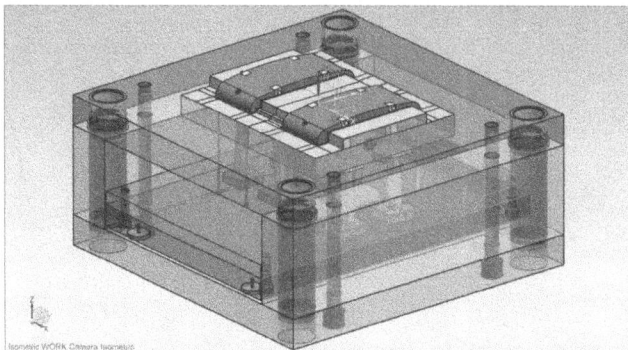

Figure 10-16 *The moving half of the mold*

Generating the Views of the Mold

Now, you need to generate the views of the mold.

1. Choose the **Assembly Drawing** tool from the **Mold Drawing** gallery of the **Mold Wizard** tab; the **Assembly Drawing** dialog box is displayed, refer to Figure 10-17. Select all the components from the drawing window and then choose the **Apply** button; the **Information** window is displayed. Choose the **Close** button to close the **Information** window.

2. Choose the **Drawing** option from the **Type** rollout and then select the **Self Contained** radio button from the **Drawing Type** rollout. Select **template_A0_asy_fam_mm.prt** from the **Templates** rollout.

3. Choose the **Apply** button and then the **Cancel** button to close the dialog box.

4. Choose the **Application** tab. Next, choose the **Drafting** tool from the **Design** group; the drafting environment is activated.

5. Choose the **Base View** tool from the **View** group; the **Base View** dialog box is displayed. Select **1:2** from the **Scale** drop-down list of the **Scale** rollout and then click to place the view, refer to Figure 10-18. Next, add dimensions to the view, refer to Figure 10-19.

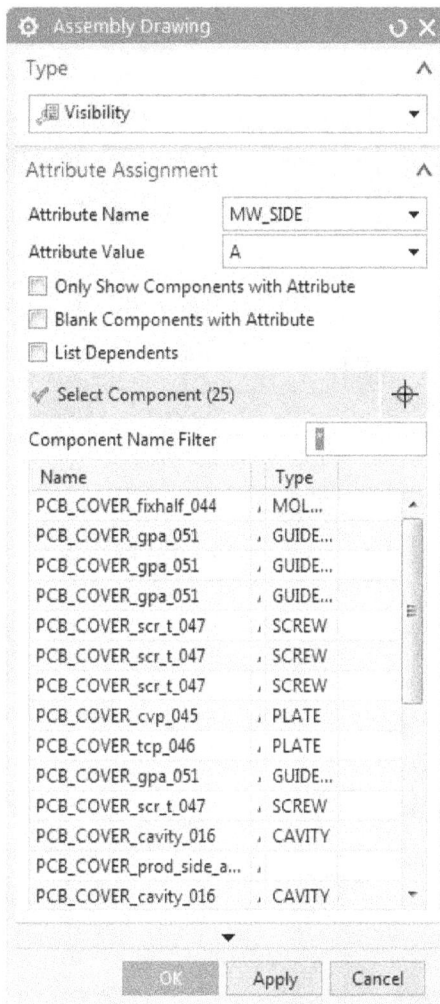

Figure 10-17 The **Assembly Drawing**
dialog box

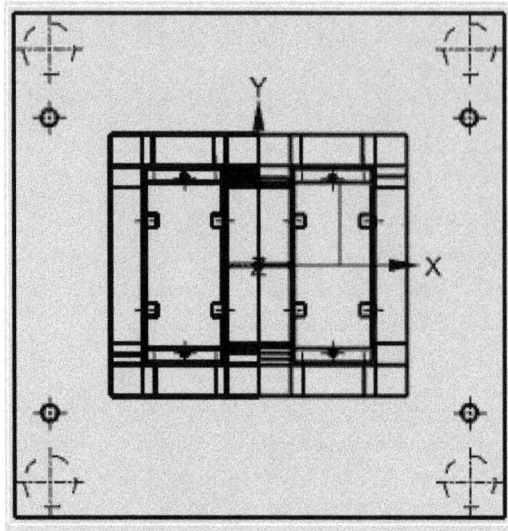

Figure 10-18 *Top view of the moving half*

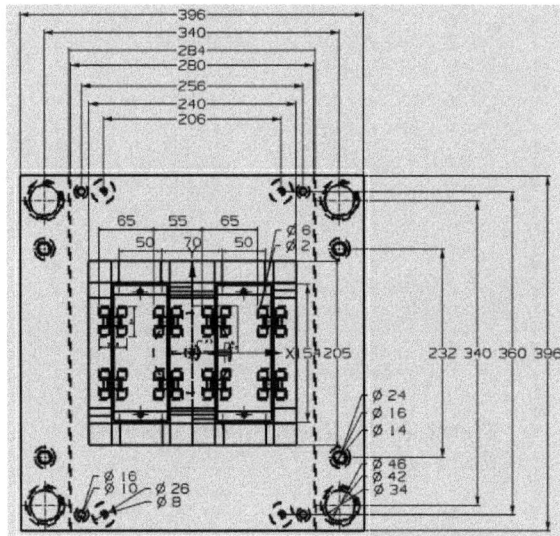

Figure 10-19 *The detailing of the moving half with hidden lines*

6. Create section line and section view, refer to Figures 10-20 and 10-21.

7. Repeat the same procedure for detailing the other views, refer to Figures 10-22 and 10-23.

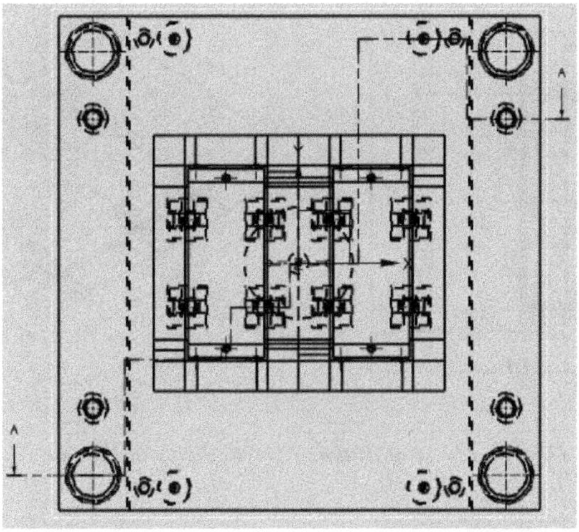

Figure 10-20 The Top view of the moving half of the mold

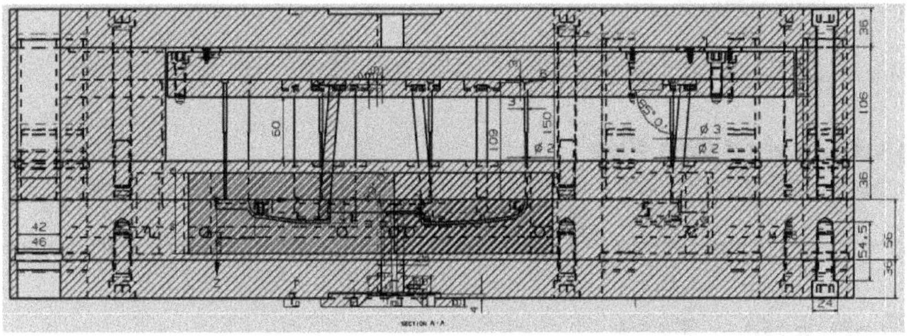

Figure 10-21 Section line on the moving half

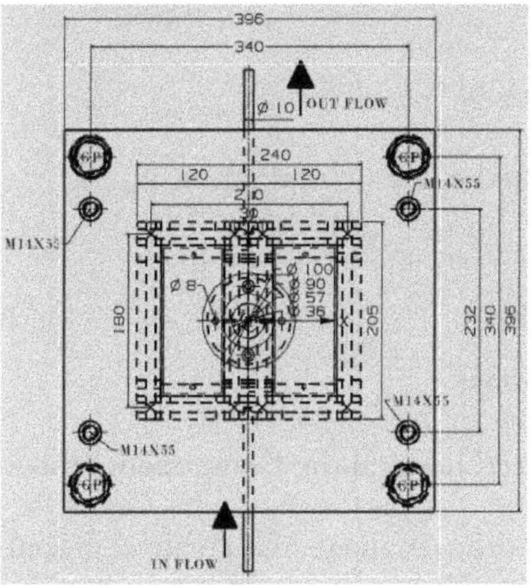

Figure 10-22 Inverted Top view of the stationary half

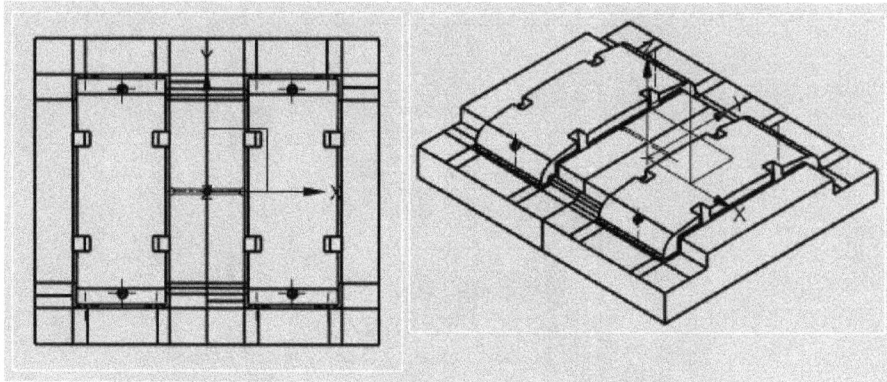

Figure 10-23 *The Top and Isometric views of the core insert*

Note
1. Before creating the section view, you need to display all the components of the mold in the mold wizard environment.

2. Add in flow and out flow annotations manually.

Saving and Closing the File

1. Choose **Menu > File > Close > Save and Close** from the **Top Border Bar** to save and close the file.

Self-Evaluation Test

Answer the following questions and then compare them to those given at the end of this chapter:

1. The detail drawing of a mold is used for _____ .

2. The _____ tool is used to create mold drawing.

Review Questions

Answer the following questions:

1. The _____ tool is used to open the **Component Drawing** dialog box.

2. The _____ tool is used to manage and create components of mold.

EXERCISES

Exercise 1

In this exercise, you will open the output file (exr_01_top_###) of Exercise 1 of Chapter 8. Create views and dimensions to the view of the mold for the model shown in Figure 10-24 and save it.

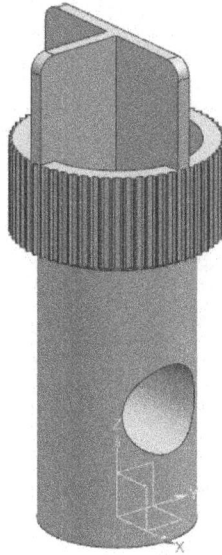

Figure 10-24 *Model for Exercise 1*

Exercise 2

In this exercise, you will open the outout file (exr_02_top_###) of Exercise 2 of Chapter 8. Create views and dimensions to the view of the mold for the model shown in Figure 10-25 and save it.

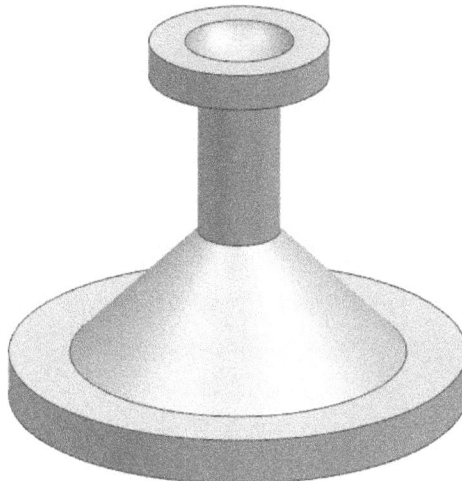

Figure 10-25 *Model for Exercise 2*

Answers to Self-Evaluation Test

1. manufacturing, **2. Assembly Drawing**

Index

Other Publications by CADCIM Technologies

The following is the list of some of the publications by CADCIM Technologies. Please visit *www.cadcim.com* for the complete listing.

AutoCAD Textbooks
- AutoCAD 2018: A Problem-Solving Approach, Basic and Intermediate, 24th Edition
- Advanced AutoCAD 2018: A Problem-Solving Approach (3D and Advanced), 24th Edition
- AutoCAD 2017: A Problem-Solving Approach, Basic and Intermediate, 23rd Edition
- AutoCAD 2017: A Problem-Solving Approach, 3D and Advanced, 23nd Edition
- AutoCAD 2016: A Problem-Solving Approach, Basic and Intermediate, 22nd Edition
- AutoCAD 2016: A Problem-Solving Approach, 3D and Advanced, 22nd Edition

Autodesk Inventor Textbooks
- Autodesk Inventor Professional 2018 for Designers, 18th Edition
- Autodesk Inventor Professional 2017 for Designers, 17th Edition
- Autodesk Inventor 2016 for Designers, 16th Edition
- Autodesk Inventor 2015 for Designers, 15th Edition

AutoCAD MEP Textbooks
- AutoCAD MEP 2018 for Designers, 4th Edition
- AutoCAD MEP 2016 for Designers, 3rd Edition
- AutoCAD MEP 2015 for Designers

Solid Edge Textbooks
- Solid Edge ST10 for Designers, 15th Edition
- Solid Edge ST9 for Designers, 14th Edition
- Solid Edge ST8 for Designers, 13th Edition

NX Textbooks
- NX 12.0 for Designers, 11th Edition
- NX 11.0 for Designers, 10th Edition
- NX 10.0 for Designers, 9th Edition

NX Nastran Textbook
- NX Nastran 9.0 for Designers

SolidWorks Textbooks
- SOLIDWORKS 2018 for Designers, 16th Edition
- SOLIDWORKS 2017 for Designers, 15th Edition
- SOLIDWORKS 2016 for Designers, 14th Edition
- SolidWorks 2014: A Tutorial Approach
- SolidWorks 2012: A Tutorial Approach
- Learning SolidWorks 2011: A Project Based Approach

SolidWorks Simulation Textbooks
• SOLIDWORKS Simulation 2018: A Tutorial Approach
• SOLIDWORKS Simulation 2016: A Tutorial Approach

CATIA Textbooks
• CATIA V5-6R2017 for Designers, 15th Edition
• CATIA V5-6R2016 for Designers, 14th Edition
• CATIA V5-6R2015 for Designers, 13th Edition

Creo Parametric and Pro/ENGINEER Textbooks
• Creo Parametric 4.0 for Designers, 4th Edition
• PTC Creo Parametric 3.0 for Designers, 3rd Edition
• Creo Parametric 2.0 for Designers
• Pro/Engineer Wildfire 5.0 for Designers
• Pro/ENGINEER Wildfire 4.0 for Designers
• Pro/ENGINEER Wildfire 3.0 for Designers

AutoCAD Plant 3D Textbooks
• AutoCAD Plant 3D 2018 for Designers, 4th Edition
• AutoCAD Plant 3D 2016 for Designers, 3rd Edition
• AutoCAD Plant 3D 2015 for Designers

ANSYS Textbooks
• ANSYS Workbench 14.0: A Tutorial Approach
• ANSYS 11.0 for Designers

Creo Direct Textbook
• Creo Direct 2.0 and Beyond for Designers

Autodesk Alias Textbooks
• Learning Autodesk Alias Design 2016, 5th Edition
• Learning Autodesk Alias Design 2015, 4th Edition
• Learning Autodesk Alias Design 2012
• AliasStudio 2009 for Designers

AutoCAD LT Textbooks
• AutoCAD LT 2017 for Designers, 12th Edition
• AutoCAD LT 2016 for Designers, 11th Edition
• AutoCAD LT 2015 for Designers, 10th Edition

EdgeCAM Textbooks
• EdgeCAM 11.0 for Manufacturers
• EdgeCAM 10.0 for Manufacturers

AutoCAD Electrical Textbooks
- AutoCAD Electrical 2018 for Electrical Control Designers, 9th Edition
- AutoCAD Electrical 2017 for Electrical Control Designers, 8th Edition
- AutoCAD Electrical 2016 for Electrical Control Designers, 7th Edition

Autodesk Revit Architecture Textbooks
- Exploring Autodesk Revit 2018 for Architecture, 14th Edition
- Exploring Autodesk Revit 2017 for Architecture, 13th Edition
- Autodesk Revit Architecture 2016 for Architects and Designers, 12th Edition

Autodesk Revit Structure Textbooks
- Exploring Autodesk Revit 2018 for Structure, 8th Edition
- Exploring Autodesk Revit 2017 for Structure, 7th Edition
- Exploring Autodesk Revit Structure 2016, 6th Edition

Autodesk Revit MEP Textbooks
- Exploring Autodesk Revit 2018 for MEP, 5th Edition
- Exploring Autodesk Revit 2017 for MEP, 4th Edition
- Exploring Autodesk Revit MEP 2016, 3rd Edition

AutoCAD Civil 3D Textbooks
- Exploring AutoCAD Civil 3D 2018, 8th Edition
- Exploring AutoCAD Civil 3D 2017, 7th Edition
- Exploring AutoCAD Civil 3D 2016, 6th Edition

AutoCAD Map 3D Textbooks
- Exploring AutoCAD Map 3D 2018, 8th Edition
- Exploring AutoCAD Map 3D 2017, 7th Edition
- Exploring AutoCAD Map 3D 2016, 6th Edition

RISA-3D Textbook
- Exploring RISA-3D 14.0

Autodesk Navisworks Textbooks
- Exploring Autodesk Navisworks 2017, 4th Edition
- Exploring Autodesk Navisworks 2016, 3rd Edition

AutoCAD Raster Design Textbooks
- Exploring AutoCAD Raster Design 2017
- Exploring AutoCAD Raster Design 2016

Bentley STAAD.Pro Textbooks
- Exploring Bentley STAAD.Pro CONNECT Edition
- Exploring Bentley STAAD.Pro V8i (SELECTseries 6)
- Exploring Bentley STAAD.Pro V8i

3ds Max Design Textbooks
• Autodesk 3ds Max Design 2015: A Tutorial Approach, 15th Edition
• Autodesk 3ds Max Design 2014: A Tutorial Approach
• Autodesk 3ds Max Design 2013: A Tutorial Approach

3ds Max Textbooks
• Autodesk 3ds Max 2018: A Comprehensive Guide, 18th Edition
• Autodesk 3ds Max 2018 for Beginners: A Tutorial Approach, 18th Edition
• Autodesk 3ds Max 2017: A Comprehensive Guide, 17th Edition
• Autodesk 3ds Max 2017 for Beginners: A Tutorial Approach, 17th Edition
• Autodesk 3ds Max 2016: A Comprehensive Guide, 16th Edition
• Autodesk 3ds Max 2016 for Beginners: A Tutorial Approach, 16th Edition

Autodesk Maya Textbooks
• Autodesk Maya 2018: A Comprehensive Guide, 10th Edition
• Autodesk Maya 2017: A Comprehensive Guide, 9th Edition
• Autodesk Maya 2016: A Comprehensive Guide, 8th Edition
• Character Animation: A Tutorial Approach

Pixologic ZBrush Textbooks
• Pixologic ZBrush 4R7: A Comprehensive Guide, 3rd Edition
• Pixologic ZBrush 4R6: A Comprehensive Guide

Fusion Textbooks
• Blackmagic Design Fusion 7 Studio: A Tutorial Approach
• The eyeon Fusion 6.3: A Tutorial Approach

Flash Textbooks
• Adobe Flash Professional CC 2015: A Tutorial Approach
• Adobe Flash Professional CC: A Tutorial Approach
• Adobe Flash Professional CS6: A Tutorial Approach

Computer Programming Textbooks
• Introducing PHP/MySQL
• Introduction to C++ programming, 2nd Edition
• Learning Oracle 12c - A PL/SQL Approach
• Learning ASP.NET AJAX
• Introduction to Java Programming, 2nd Edition
• Learning Visual Basic.NET 2008

MAXON CINEMA 4D Textbook
• MAXON CINEMA 4D R18 Studio: A Tutorial Approach, 5th Edition

Oracle Primavera Textbook
• Exploring Oracle Primavera P6 v8.4

AutoCAD Textbooks Authored by Prof. Sham Tickoo and Published by Autodesk Press
• AutoCAD: A Problem-Solving Approach: 2013 and Beyond
• AutoCAD 2012: A Problem-Solving Approach
• AutoCAD 2011: A Problem-Solving Approach
• AutoCAD 2010: A Problem-Solving Approach
• Customizing AutoCAD 2010
• AutoCAD 2009: A Problem-Solving Approach

Textbooks Authored by CADCIM Technologies and Published by Other Publishers

3D Studio MAX and VIZ Textbooks
• Learning 3DS Max: A Tutorial Approach, Release 4
  Goodheart-Wilcox Publishers (USA)
• Learning 3D Studio VIZ: A Tutorial Approach
  Goodheart-Wilcox Publishers (USA)

CADCIM Technologies Textbooks Translated in Other Languages

SolidWorks Textbooks
• SolidWorks 2008 for Designers (Serbian Edition)
  Mikro Knjiga Publishing Company, Serbia
• SolidWorks 2006 for Designers (Russian Edition)
  Piter Publishing Press, Russia
• SolidWorks 2006 for Designers (Serbian Edition)
  Mikro Knjiga Publishing Company, Serbia

NX Textbooks
• NX 6 for Designers (Korean Edition)
  Onsolutions, South Korea
• NX 5 for Designers (Korean Edition)
  Onsolutions, South Korea

Pro/ENGINEER Textbooks
• Pro/ENGINEER Wildfire 4.0 for Designers (Korean Edition)
  HongReung Science Publishing Company, South Korea
• Pro/ENGINEER Wildfire 3.0 for Designers (Korean Edition)
  HongReung Science Publishing Company, South Korea

Autodesk 3ds Max Textbook
• 3ds Max 2008: A Comprehensive Guide (Serbian Edition)
  Mikro Knjiga Publishing Company, Serbia

AutoCAD Textbooks
• AutoCAD 2006 (Russian Edition)
  Piter Publishing Press, Russia
• AutoCAD 2005 (Russian Edition)
  Piter Publishing Press, Russia
• AutoCAD 2000 Fondamenti (Italian Edition)

Coming Soon from CADCIM Technologies
• SolidCAM 2016: A Tutorial Approach
• Autodesk Fusion 360: A Tutorial Approach
• Project Management Using Microsoft Project 2016 for Project Managers

Online Training Program Offered by CADCIM Technologies
CADCIM Technologies provides effective and affordable virtual online training on animation, architecture, and GIS softwares, computer programming languages, and Computer Aided Design, Manufacturing, and Engineering (CAD/CAM/CAE) software packages. The training will be delivered 'live' via Internet at any time, any place, and at any pace to individuals, students of colleges, universities, and CAD/CAM/CAE training centers. For more information, please visit the following link: *http://www.cadcim.com*.

www.ingramcontent.com/pod-product-compliance
Lightning Source LLC
Chambersburg PA
CBHW080525220326
41599CB00032B/6199